The changing climate:
responses of the natural fauna and flora

The changing climate: responses of the natural fauna and flora

MICHAEL J. FORD

London
GEORGE ALLEN AND UNWIN
Boston   Sydney

George Allen & Unwin (Publishers) Ltd,
40 Museum Street, London WC1A 1LU, UK

George Allen & Unwin (Publishers) Ltd,
Park Lane, Hemel Hempstead, Herts HP2 4TE, UK

Allen & Unwin Inc.,
9 Winchester Terrace, Winchester, Mass 01890, USA

George Allen & Unwin Australia Pty Ltd,
8 Napier Street, North Sydney, NSW 2060, Australia

First published in 1982

British Library Cataloguing in Publication Data

Ford, Michael J.
    The changing climate.
1. Bioclimatology 2. Climatic changes
I. Title
574.5'222   QH543
ISBN 0-04-574017-8

Library of Congress Cataloging in Publication Data

Ford, Michael J. (Michael John), 1950–
    The changing climate.
Includes bibliographical references and index.
1. Bioclimatology. 2. Acclimatization. 3. Climatic changes. I. Title.
QH543.F67   574.5'222   81-20065
ISBN 0-04-574017-8   AACR2

Set in 10 on 12 point Times by Gilbert Composing, Leighton Buzzard,
and printed in Great Britain
by Richard Clay (The Chaucer Press) Ltd, Bungay, Suffolk.

Preface

I believe that this book is the first attempt that has been made to present a comprehensive account of the various ways that animals and plants occurring in natural habitats respond to the changing climate. I have made no attempt to deal with productive ecosystems, partly because there is already a substantial literature covering the fields of agricultural meteorology and the effects of climate on crops, but also because it is clear that the responses of agricultural species to climate are atypical in that they can be directly manipulated by man, in particular by the artificial selection of tolerant strains.

It is clear that the climatic changes that have occurred in recent times, although significant, are of relatively small magnitude in comparison with those that have occurred since the end of the last ice age some 10 000 years ago; however, as I hope this book will show, the natural fauna and flora have always been dependent upon climatically favourable refuges from which to re-disperse after climatic deteriorations, whether these be unfavourable trends or individual extremes such as a severe winter. With the environmental impact of man's activities now resulting in the progressive destruction and fragmentation of the last remaining areas of natural habitat, the additional pressure of a climatic extreme may threaten a species with local extinction. An examination of the climatic responses of species, especially those at the edge of their range, is thus a prerequisite for their successful conservation.

London, 1981   MICHAEL J. FORD

Acknowledgements

I would like to thank the following individuals and organisations for permission to reproduce illustrative material (numbers in parentheses refer to text figures):

H. H. Lamb (1.1, 3.1, 3.2, 3.3, 3.5, 4.1, 4.2); H. Godwin (2.1); C. D. Pigott (2.2, 6.4); P. D. Jones (3.4); R. Spencer (3.6); Figure 3.7 reprinted by permission from *Nature*, Vol. 245, pp. 190–4, Copyright © 1973 Macmillan Journals Ltd; Royal Meteorological Society (4.3, 6.5, 6.7, 7.4); C. R. Tubbs, C. J. Bibby and the Editor, *British Birds* (5.1); J. E. Satchell (5.2, 8.1); E. B. Ford and Collins Ltd (5.3, 5.4); the General Secretary, Institute of Biology (6.1); J. W. King (6.3); J. A. Kington (6.6); F. H. Perring (7.2); Figure 7.3 reproduced by permission of the Controller, Her Majesty's Stationery Office, Crown copyright reserved; Clare S. Lloyd (8.2); J. Stafford (9.1); the Editor, *Journal of Ecology* (10.1); A. J. Southward and D. J. Crisp (10.2); C. B. Cox, P. D. Moore and Blackwell Scientific Publications Ltd (11.1).

Contents

List of tables

1
Introduction

Climate changes on all time-scales from the gusting of the wind to the periodic occurrence of series of ice ages; but one of the most significant of recent climatic changes – probably unprecedented in the last 1000 years – has been the remarkable warming experienced during the early part of the present century. The first half of the twentieth century was, on a global scale, a period of equable climate with fewer extremes of heat and cold than at present and with more consistent rainfall penetrating into the continental interiors. However, in the northern hemisphere at least, this amelioration ended in the late 1940s (twentieth century climatic changes are examined in more detail in Ch. 3), since when temperatures have dropped. Furthermore, this deterioration appears to have been accompanied by an increase in weather variability with a greater frequency of extremes of heat and cold, flood and drought.

Although, because conditions in general became more favourable for man and his activities, the twentieth century climatic amelioration was only recognised retrospectively, the subsequent deterioration was registered much more rapidly, primarily because of its effect on food production around the world. The increased variability in tropical rainfall led to harvest failures and consequent famines such as that experienced in the Sahel region of Africa between 1968 and 1973 (see Sec. 3.3), but even in temperate regions the risk of harvest failure has increased after the optimal conditions that prevailed in the 1940s and 1950s. As an example, in 1972 a combination of severe winter weather but an absence of protective snow cover followed by a summer drought resulted in a 12% shortfall in the Russian grain harvest. In consequence the USSR purchased 28 million tonnes of North American grain, which resulted in the price of wheat in western countries doubling in 4 months. The world's reserves of grain throughout the 1970s were generally about one-fifth of the levels maintained during the 1960s and, should harvest failures occur simultaneously in more than one of the world's great grain-growing regions, the consequences in terms of widespread food shortages – and even mass starvation – would be formidable: climate is the factor which will determine at what date Malthus' (1766–1834) prediction that the human population will eventually outstrip its food resources will be fulfilled. It is by its effect on food production that climatic change will impinge directly on man and his survival, and this is a much more immediate threat than, for example, speculations on the imminence of a future ice age.

But what of the future trend of the changing climate? The science of climatology is in its infancy and the mechanisms that bring about climatic changes are inadequately known, although some are discussed in Section 3.4. In consequence, any predictions of future climatic conditions have to be made either by searching for past analogues of present conditions and then extrapolating their consequences, or on the basis of statistical relationships, the physical mechanisms of which are unknown. On the evidence that we have available at the present time it seems likely that a maintenance of the recent cooling trend would be the natural course of events for the forseeable future. However, man is just beginning to have a significant influence on the global climate, notably via the release of carbon dioxide as a result of the burning of fossil fuel, and this 'anthropogenic' influence is tending to bring about a warming of the atmosphere by means of the 'greenhouse effect' discussed in Section 3.4. In the present state of knowledge it is not possible to predict the outcome of these two conflicting trends – the natural and the anthropogenic – although even if climatic conditions remained constant the significance of their impact on man would become progressively greater as the human population continues to increase.

The responses of the natural fauna and flora to climatic changes that are described in this book are considered against the background of twentieth century climatic *trends;* but, as described above, temperature and rainfall trends, identified over decades, are accompanied by variations in the frequency of individual weather events, particularly meteorological extremes. The distinction between weather and climate is thus not clear, although in general climate constitutes the long-term summation of the weather at a particular site. Biological organisms may respond to a wide range of weather variables, as described in Section 2.2, and are frequently sensitive to the totality of climatic effects. As described in Section 3.2, changes in the readily quantifiable meteorological parameters such as temperature and rainfall are merely indicators of changes in the strength and location of large scale features of the atmospheric circulation. In consequence any index of the circulation, such as that devised by H.H. Lamb (1972a) (see Sec. 3.2), which summates the influence of a number of climatic variables has considerable biological significance. The Lamb index is an index of the strength of the westerly winds, the main feature of the atmospheric circulation in middle latitudes, and hence an indication of the oceanicity of the climate. As described in Section 3.2, when the westerly circulation is vigorous the moderating influence on the climate exerted by the oceans extends further eastwards into the continental interiors: the air masses that reach the western shores of the continents have generally travelled across long stretches of water which, owing to its high specific heat (the specific heat is the amount of heat required to change the temperature of 1 g of a substance by 1°C), heats up and cools down slowly. This means that in winter the sea remains relatively warm in comparison with the land and so heats up the air passing over it, whilst in summer it remains relatively cold with respect to the land and so cools down the overlying air stream. The result is a general amelioration of the temperature régime of maritime or oceanic regions such that their winters are less cold and their summers less warm than they would otherwise be. As the airstream continues inland, this moderating influence is no longer effective and so continental regions experience colder winters and warmer summers than oceanic regions on the same latitude (see Fig. 1.1). The westerly winds

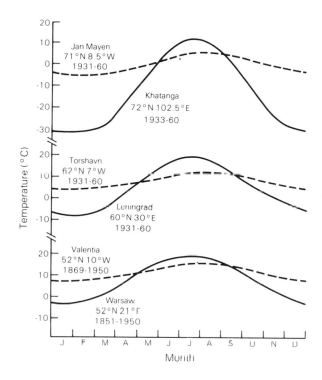

Figure 1.1 Annual cycle of temperature at oceanic (----) and continental (—) sites at similar latitudes. Redrawn from Lamb (in press).

also pick up moisture over the oceans and carry with them rain-producing depressions. In consequence maritime areas are characterised by higher rainfall than continental interiors, and this rainfall is generally well distributed throughout the year in contrast to the marked rainy seasons that occur further inland.

Although the use of the terms 'oceanic' (or 'maritime') and 'continental' to describe prevailing or changing climatic conditions is not favoured by climatologists – who in general prefer the precision of quantitative instrumental data on temperature, rainfall, wind speed and direction, etc. – these terms are valid when considering biological organisms which respond to the interacting totality of meteorological parameters. As described in Chapter 3, the climatic trends experienced in middle latitudes during the present century may be described as indicating a change from an increased predominance of oceanic influence in the first part of the century to a more continental régime since the 1940s. The location of the British Isles at the eastern limit of a large ocean means that any changes in the strength of the westerly circulation are quickly registered as changes in the oceanicity of the British climate. Thus the British Isles constitute a particularly sensitive area in which climatic changes and their effects are indicative of the trends prevailing throughout middle latitudes and thus, via the kind of teleconnections discussed in Section 3.3, the global climate as a whole.

When one starts to consider the various ways in which wild plants and animals respond to climate and climatic change, it becomes clear that although the primary or direct effects of climate tend to be physiological, as described in Chapter 6, there are a number of indirect climatic effects on the ecology of organisms. For example, the changes in the atmospheric circulation that are the driving force behind the more readily quantifiable changes in climatic variables may transport organisms to new, and perhaps more favourable, environments (Ch. 7). Furthermore, some species may indirectly register the direct effects of climate on other organisms, whether pathogens (Ch. 8) or prey (Ch. 9). Climate may also influence the outcome of competitive interactions: plants and animals seldom occur as extensively as their tolerance limits to climatic parameters would permit, but are usually limited by competition with other species. As climate changes, the balance of competitive advantage shifts and readjustments of distributional limits ensue (Ch. 10).

However, climate constitutes only one of the environmental factors influencing the survival of an organism and other features of the habitat interact with, and modify the influence of, climatic fluctuations. This interaction is discussed in Chapter 5, after a consideration of the nature of recent (Ch. 3) and past (Ch. 4) climatic changes. The importance of such habitat features may not be manifest during climatically favourable periods, but even quite widely distributed species may retreat to optimal areas when the climate deteriorates. If other environmental pressures, notably those associated with the activities of man, have diminished or even destroyed these 'refuges' during the intervening period, then the survival of these species may be endangered. We may be unable to do anything to modify climatic trends directly, but we can ameliorate their effects on our natural fauna and flora by safeguarding key sites and attempting to reduce the impact of other, more controllable, pressures on the environment.

2

Biological principles relating to the action of climatic factors

In order to comprehend fully the diverse ways in which organisms respond to the changing climate it is necessary to consider some of the fundamental physiological and ecological principles governing the biology of animals and plants, and the way in which these interact with climatic factors.

2.1 Temperature and living organisms

Climatic changes, on a variety of time-scales, are generally identified on the basis of temperature trends, although they usually entail changes in other meteorological variables, such as rainfall, windiness and duration of sunshine (see Sec. 2.2). Temperature is probably the most significant climatic factor in biological terms as all metabolic processes (indeed most chemical reactions) are temperature-dependent. The processes of photosynthesis, respiration, digestion, excretion and also growth and activity essentially obey Van't Hoff's law, which states that the logarithm of the velocity of a given reaction is proportional to the temperature at which it takes place. This relationship is usually expressed by means of a temperature coefficient, or Q_{10}, indicating the increase in the rate of a process produced by raising the temperature by 10°C, and which may be calculated from the formula

$$ Q_{10} = \left(\frac{M_2}{M_1} \right)^{10/(t_2 - t_1)} $$

where M_1 is the rate at temperature t_1 and M_2 is the rate at temperature t_2. The earlier generalization that the Q_{10} for biological reactions is always about 2 (Krogh 1914) is no longer tenable as values up to 5 (e.g. Berthet 1964) or more have been demonstrated. Q_{10} values are generally higher at lower temperatures and Rao and Bullock (1954) have shown an increase in Q_{10} with body weight in a number of invertebrates.

All physiological processes are catalysed by enzymes which, in common with other proteins, are inactivated or denatured at high temperatures. The enzyme complex of an organism has evolved so that it functions most efficiently at an optimal temperature which is appropriate to the prevailing temperatures in the organism's habitat (in the case of mammals and birds the optimum temperature is the regulated body temperature somewhere between 30 and 40°C). This explains why apparently more favourable temperatures may be deleterious to an organism which is not adapted to them. An extreme example of this is the case of the fish *Trematomus bernacchi* which is found in cold Antarctic waters of about 0°C and becomes immobile with heat prostration if temperatures approach 2°C (Wohlschlag 1960). In addition to the long-term evolution of such enzymically determined tolerance limits, individual animals and plants may exhibit short-term adaptation or acclimatisation which involves changes in the enzyme–substrate relationships at the cellular level (Somero 1969). An analogous process is the 'hardening' of plants at low temperatures, which involves a redistribution of water in the plant, enabling it ultimately to withstand much lower temperatures provided that these are attained gradually.

The phenomenon of warm-bloodedness, which enables animals to be relatively independent of prevailing temperatures, is a comparatively recent evolutionary development which is now found only in birds and mammals but perhaps formerly also occurred in some dinosaurs (Bakker 1971, 1972). In contrast, cold-blooded animals and all plants are much more directly affected by ambient temperatures and thus are more susceptible to climatic changes. However, the energy costs of warm-bloodedness increase as temperatures drop and so the metabolism of mammals and birds is still directly affected by temperature.

The terms 'homeothermic', or 'homoiothermic', and 'poikilothermic' are effectively equivalent to 'warm-blooded' and 'cold-blooded', respectively, but the former terminology emphasises the constancy or otherwise of body temperature. Thus, homeothermic animals are those which maintain a stable body temperature by means of a number of homeostatic mechanisms such as shivering, sweating and panting which are controlled by the hypothalamic region of the brain. The precise temperature that is maintained varies from species to species but is generally higher in birds (38–42°C) than mammals (35–39°C). In contrast, the body temperature of poikilotherms is not stable but varies over a considerable range. It does not necessarily follow changes in ambient temperature as poikilotherms are also able to regulate their body temperature to some extent. The main difference between homeothermic and poikilothermic thermoregulation (Whittow 1970, 1971) is that in the former the source of heat is mainly internal (i.e. endothermic) in that it is the result of the animal's accelerated metabolism. Although poikilotherms can produce some endothermic heat by such mechanisms as shivering or, in the case of insects, the production of muscle heat by wing vibration, the main source of heat is the external environment, i.e. these organisms are ectothermic. Thermoregulation in poikilotherms usually involves the alternation of periods of basking and sheltering in the shade, or changing the orientation of the body to the sun.

2.2 Other climatic factors

Although temperature is the climatic factor of greatest significance in biological terms, other aspects of climate are also important to living things. Thermoregulation by basking has already been mentioned above and this leads one to distinguish between the effects of sunshine and those of air temperature. By directly absorbing radiant energy, animals may become mobile even if air temperatures are below those necessary for normal activity. This may account for the fact that many ground-living invertebrates (e.g. beetles and spiders) are dark-coloured and so absorb a greater amount of incoming solar radiation.

The distribution of a number of open ground-nesting ants appears to be determined by a requirement for insolation (Felton 1974) and the woodland species *Formica rufa* is confined to sites with more than 40 days a year with greater than 9 h bright sunshine (Hughes 1975). The extinction of a number of colonies of this species near its northern limit in the Lake District has been attributed (Satchell & Collingwood 1955) to a decline in the amount of solar radiation reaching the ground, resulting from the invasion of the woods by bracken (*Pteridium aquilinum*). Similarly many reptiles bask in the sun (they may be called 'heliotherms' – see Cloudsley-Thompson 1971) and it has been suggested (Jackson 1978) that a reduction in the duration of May sunshine during the 1960s may have been a possible factor in the decline of the sand lizard (*Lacerta agilis*) on the edge of its range in north-western England.

In plants, sunshine is clearly vital for photosynthesis and hence growth, but it is unlikely to be a limiting factor apart from in competitive interactions where large plants shade small ones. Owing to their lack of mobility, plants are more susceptible to the deleterious effects of an excess of sunshine than are animals. An example is the phenomenon of sun-scald in thin barked trees such as ash (*Fraxinus excelsior*), aspen (*Populus tremula*) and birch (*Betula pendula*). This involves the death of a long strip of bark on one side of the trunk and may be due either to absolute overheating (e.g. Rackham 1975 recorded a temperature of 46.5°C on the surface of an alder (*Alnus glutinosa*) pole exposed to unaccustomed direct sunlight at the edge of a coppiced area) or to a sudden thawing when the Sun's rays strike a frozen trunk.

The importance of moisture to plants and animals is manifest when one considers that all living things are composed of about 80% water and that protoplasm is only physiologically active when fully hydrated. Many seeds and spores (of both animals and plants) can exist in a dormant state until moisture conditions are suitable to re-activate their protoplasm. A similar phenomenon involving whole adult organisms occurs in certain lower plants, such as algae, fungi, lichens and even some bryophytes and pteridophytes, which may be termed poikilohydric (an equivalent term to poikilothermic). In these plants the degree of hydration depends on the humidity of the surrounding air; if this drops the organisms pass into a latent or resting state only to revive again when suitable conditions of moisture return. A few animals, such as tardigrades and bdelloid rotifers, which live in the water films surrounding terrestrial mosses and lichens, have evolved a similar response to dry conditions which is termed anabiosis or cryptobiosis. These animals can withstand extreme desiccation: when the ambient humidity is low they dry out and contract; when water is present again the animals swell and become active once more, within a few hours.

The relative humidity of the atmosphere and the effectiveness of precipitation are directly affected by prevailing temperatures and most classifications of regional climates, especially those based on vegetation type (Köppen 1923, 1931, Thornthwaite 1933, 1948), recognise this interaction between temperature and moisture. Thus, the climate of an area may be effectively represented as a climograph of temperature plotted against humidity or rainfall month by month to produce a twelve-sided polygon.

When considering climatic changes one can distinguish trends in rainfall and also variations in the frequency of extremes, namely floods and droughts. Animals, being mobile, are often able to move from areas of unusually high or low rainfall to a more hospitable environment. Clearly relative mobility is dependent upon the size of the organism concerned; thus Varley (1947) showed that summer floods drowned a proportion of the larvae, as well as the pupae, of the knapweed gallfly (*Urophora jaceana*). Eggs, of course, are not mobile and Dempster (1971) found that heavy rain knocked eggs of the small copper (*Lycaena phlaeas*) off the food plant sheep's sorrel (*Rumex acetosella*). In such a situation the eggs may fail to hatch or may be rendered more accessible to predators. Rain may be a significant source of mortality in aphids: Dixon (1976) has demonstrated high mortality attributable to rain in the sycamore aphid (*Drepanosiphum platanoides*) where, if egg hatching is not closely synchronised with bud burst in the host tree *Acer pseudoplatanus,* aphids exposed on the smooth sycamore buds may be washed off the tree. Rain has also been shown to be one of the most consistent factors reducing populations of the pea aphid (*Acyrthosiphon pisum*) in East Anglia (Dunn & Wright 1955). The young of homeothermic species may also suffer mortality associated with rainfall causing loss of body heat with consequent death by exposure. This is a significant cause of fledgling mortality in ground-nesting birds such as the stone curlew (*Burhinus oedicnemus*).

In plants, apart from the physical damage caused by heavy rainfall, the soil may become waterlogged, and this results in anoxic conditions which reduce root respiration. Additionally, soil nutrients and even the topsoil itself may be washed away. Extreme drought, as in 1976 in England (see Hearn & Gilbert 1977) may cause the death of plants, including mature trees. Again owing to their mobility, animals suffer less from drought, but this does not, of course, apply to animals restricted to aquatic habitats for all or part of their life-cycle. The drying up of breeding ponds reduces the reproductive success of amphibians in hot summers, notably so in the case of the natterjack toad (*Bufo calamita*) (see Sec. 5.3), whose eggs require high temperatures for development and so are laid in shallow pools which heat up rapidly but in consequence are prone to desiccation. Similarly, in drought years river levels drop and fish and aquatic invertebrates suffer the combined effects of high temperatures and low oxygen levels, the latter partly attributable to algal blooms. There were many reports of deaths of salmon (*Salmo salar*) in England in the hot summers of 1975 and 1976 and a study of one example of a mass mortality of this fish in the lower reaches of the River Wye in late June 1976 (Brooker, Morris & Hemsworth 1977) indicated that whilst high water temperatures (up to 27.6°C) were a contributory factor, the main cause of the mortality was the low oxygen concentration resulting from the decay of large quantities of the submerged macrophyte river crowfoot (*Ranunculus fluitans*) which had grown prolifically in what were optimal temperature conditions.

On land, egg-laying in the snail *Helix pomatia* is very dependent upon rainfall as the soil in which the eggs are placed must be thoroughly wet to the depth of the egg cavity (about 6 cm) before the eggs are laid. This is not just an effect of rainfall upon digging activity as on a number of occasions a cavity is dug but eggs are not laid, presumably because the ground at the base of the cavity is found to be insufficiently moist. Short periods of heavy rain, which wet the soil to a considerable depth, provide more suitable conditions for egg-laying than longer spells of light rain (Pollard 1975). There was virtually no breeding of this snail in 1976 when, owing to the drought, the ground was not sufficiently wet to induce egg-laying. Similarly, the breeding of birds such as swallows (*Hirundo rustica*) and martins (*Delichon urbica*) is less successful in drought years when mud for nest-building is unavailable.

Precipitation in the form of snow has clear deleterious effects on a number of organisms, particularly by reducing the availability of prey, as in the case of the Dartford warbler (*Sylvia undata*) (see Sec. 9.2); but the insulation provided by a cover of snow is important in protecting vegetation from frost damage. Winter snow cover is particularly beneficial to heath communities which are essentially maritime or oceanic: thus dwarf shrub heath occurs up to 1100 m in Britain but reaches 2500 m in central Europe due to the longer duration of the snow cover in more continental situations (Gimingham 1972). Another oceanic species, the crowberry (*Empetrum nigrum*) (Bell & Tallis 1973), is similarly unable to survive severe winter temperatures without the insulation provided by a covering of snow but, as it flowers early in the year, it is unable to tolerate the prolongation of snow cover into the spring.

Ice cover on lakes, if prolonged, may cause oxygen depletion leading to winter kills of fish (Barica & Mathias 1979). Ice cover also provides an unusual example of a transport effect of climate which also constitutes a classic and frequently quoted example of the way in which a predator may regulate the population of its prey (Mech 1966). This relates to the population of moose (*Alces alces*) that inhabits the island of Isle Royal in Lake Superior, Canada. In the absence of any control by predators or culling by hunters this population increased to such an extent that in the 1930s the vegetation cover of the island was becoming overbrowsed and the animals were suffering from food shortage. However, in the late 1940s wolves (*Canis lupus*) crossed the frozen lake from Ontario and began to exert a predation pressure on the moose population whose numbers dropped to about 600, around which level they have stabilised ever since.

Other, less obvious, climatic factors may be significant to particular species in certain circumstances. For example, the production of toxic 'red tides' by some species of dinoflagellates appears to be dependent upon a period of calm sunny weather which allows large populations of these unicellular organisms to build up without being dispersed by the action of wind on the surface waters. The combination of wind and wave action may be particularly deleterious to those species inhabiting shorelines and populations of some shore plants are subject to periodic extinction by storm damage: thus, the sea pea (*Lathyrus japonicus*) is a very locally distributed species which occasionally appears at new coastal sites whilst disappearing from others, mainly as a result of sea storms (Perring 1974). Animals may also be similarly affected: the 1953 North Sea floods caused the extinction of the sole English colony of the damselfly *Coenagrion scitulum* which had become established at a low-lying site in Essex (Corbet, Longfield & Moore 1960) and Blus, Prouby and Neely (1979) have shown how tidal flooding is the major cause of nesting failure in the mixed colonies of royal and sandwich terns (*Sterna maxima* and *S. sandvicensis*) on the barrier islands off the coast of South Carolina.

Returning to the effects of wind action *per se,* examples of these are naturally more frequently observed in truly terrestrial habitats where plants, for example, may become structurally modified if they grow in particularly exposed situations. Newly emerged insects are especially susceptible to the effects of wind and on occasion their wings may be so damaged as to cause death, as in the case of some dragonflies (e.g. Tiensuu 1934). A particularly insidious effect of wind which is now having significant impact on certain habitats in higher latitudes is the transport of 'acid rain'. This originates from the liberation of sulphur into the atmosphere when fossil fuels such as coal and oil are burnt. The sulphur is carried eastwards on the prevailing westerly airstream (see Sec. 3.2) and in due course, for example over the eastern USA, downstream from emission sources in the Mid-West, or where the air masses rise over a line of mountains such as those fringing the coast of Norway, falls as acid rain (i.e. dilute sulphuric acid) far from its origin (Likens, Wright, Galloway & Butler 1979). Thus the United Kingdom, for example, exports its sulphur pollution to Scandinavia where freshwater habitats have been particularly affected, with significant declines in the fish populations of lakes and rivers (Leivested & Muniz 1976).

The purpose of this section has been to give an indication of the great variety of ways in which climatic factors may influence the lives of animals and plants. It is clear that in Man at least (see Sargent & Tromp 1964; Tromp 1963) there are some as yet unidentified mechanisms by which features of the climate affect the human physiology, especially in the context of meteorotropic diseases, i.e. diseases caused by or considerably affected by changes in the atmospheric environment. Thus the incidence of epileptic, asthmatic and coronary attacks appear to be correlated with the approach of cold fronts, possibly associated with some electrical disturbance propagated ahead of the front. Little progress has been made in the elucidation of such climatic influences, but, whatever the mechanisms involved, they are likely to exert a similar effect upon other animal species.

2.3   Climate and population number

Apart from the work of a handful of plant ecologists such as Harper (1967, 1977), the mechanisms that control plant population number have been rather neglected by biologists. In contrast, the control of animal populations has received great attention (see, for example, Nicholson 1933, 1954, Solomon 1949, Lack 1954a, 1966, Andrewartha & Birch 1954; Wynne Edwards 1962, Southern 1970) and has been a source of some contention and controversy amongst ecologists.

Animal populations generally fluctuate within relatively small limits about a level which is regarded as expressing the carrying capacity of the environment. It is the general consensus that such stability in animal numbers must be the result of the action of mortality factors which operate in a density-dependent way, i.e. factors which affect an increasing proportion of the population as density increases. Only factors operating in this fashion can produce the negative feedback necessary to bring about population stability. Density-dependent factors begin to act well below the carrying capacity and intensify as the population approaches an upper limit. Such mortality factors generally operate through competition for some resource that is in limited supply, such as food. As the population increases above the carrying capacity, competition becomes more intense and the resource becomes reduced, resulting in increased mortality. As the population decreases in consequence, the resource becomes relatively more abundant, competition is reduced and the mortality pressure declines. In this fashion the population oscillates around a median level, giving relative stability.

Although climate may be an important source of mortality either directly or indirectly via its effects on food supply or the incidence of disease, it always operates in a density-independent manner, irrespective of the density of the population. Thus climate can not *regulate* a population about a stable level but may *control* the numbers of a population in a non-equilibrium fashion.

The importance of density-dependent mortality factors in population regulation has been argued in great detail by David Lack (1954a, 1966), but a contrary view has been taken by Andrewartha and Birch (1954), largely on the basis of some work which apparently demonstrates that weather is the most important factor in controlling the abundance of the insect *Thrips imaginis* (Thysanoptera) (Davidson & Andrewartha 1948a, b). For 14 years the abundance of this insect on rose trees in Australian gardens was determined. Each year population numbers peaked towards the end of November (i.e. in mid-summer) and Davidson and Andrewartha were able to explain up to 84% of the variance of the size of the population peak from changes in weather factors. From this they deduced that it was not necessary to invoke density-dependent factors in order to explain the control of population numbers.

The apparent conflict of opinion between Andrewartha and Birch and their supporters and Lack and his supporters is primarily one of difference of emphasis. The former concentrate on the year-to-year differences in population numbers and are impressed by the importance of weather in determining abundance. Lack and his supporters accept that one can correlate annual fluctuations in numbers with climatic factors but they are primarily concerned with what determines the level about which such fluctuations occur. This difference of viewpoint is probably due in part to the nature of the animals under investigation. Lack studied birds, which are warm-blooded and generally survive to reproduce over several seasons and thus are less vulnerable to the climatic vagaries of a particular year. In contrast, owing to their cold-bloodedness and small individual size, insects are much more susceptible to climatic influences. They generally complete their life-cycle in a single year and sometimes have several generations per year, the actual number being determined by weather factors, and so there are considerable fluctuations, often over several orders of magnitude, in population numbers from one year to the next, although the level about which these fluctuations occur is relatively consistent. Even in those birds where there is a clear and direct climatic influence on numbers, such as the heron (see Sec. 9.2), population fluctuations are of rather small amplitude. One may similarly contrast the relative importance of climatic factors in controlling population numbers in perennial and annual plants.

One of the main problems in determining the impact of climatic, and other, factors on animal numbers is the lack of long series of annual population data. There are a few species, for example the populations of the great tit (*Parus major*) (Lack 1966), tawny owl (*Strix aluco*) (Southern 1970), and winter moth (*Operophtera brumata*) (Varley & Gradwell 1968), in Wytham Woods, Oxfordshire, which have been studied continuously for many years, but in general long runs of data are rare. In order to overcome this problem, a number of programmes of standardised population counts have been organised.

In Britain, the British Trust for Ornithology initiated an annual Common Bird Census (Williamson & Homes 1964, Williamson 1970) in 1962. As bird populations were atypically low in the years immediately following the extreme winter of 1962/3, the annual index of the Common Bird Census expresses bird populations as a percentage of that recorded in 1966, by which year it was considered that most bird species would have recovered to what might be considered a 'normal' population level. The Common Bird Census is a breeding census of birds on a nationwide sample of plots on farmland and woodland and is based on evidence of territory-holding, primarily the presence of a singing male, but other evidence, such as the finding of a nest, is acceptable. The technique has now been adopted as standard practice by the International Bird Census Committee (1969). An analogous North American Breeding Bird Survey was initiated in the eastern part of the continent in 1966 by the US Fish and Wildlife Service in collaboration with the Canadian Wildlife Service (Robbins & Van Velzen 1974, Bystrak & Robbins 1977) but was soon expanded to cover all North America north of Mexico. It is based on standard transects along roadside routes during the breeding season.

Transects have also been used to assess the annual abundance of butterflies in parts of Britain (Moore 1975, Pollard, Elias, Skelton & Thomas 1975) and in 1976 a national butterfly monitoring scheme was set up (Pollard 1977, 1979a). It has long been assumed that butterfly numbers are in some way related to weather conditions and Beirne (1947) attempted a more sophisticated analysis of this relationship by examining past accounts of 'good' and 'bad' years for Lepidoptera between 1865 and 1944 and comparing these with the available climatic information. He concluded that: a severe and prolonged frost in winter is generally favourable as it kills off large numbers of insectivorous birds, although it may be detrimental to larvae; late spring or early autumn frosts are always deleterious; mild winters are unfavourable because natural enemies, including pathogens, flourish in mild conditions; high rainfall is detrimental and Lepidoptera are generally exceptionally abundant in long, warm, dry summers, although excessive summer drought is also deleterious. Beirne's data indicate a decline in abundance of butterflies through the 1920s when there were a number of successive poor breeding seasons due to high rainfall, followed by a period of expansion in the good seasons of the 1930s.

Ekholm (1975) studied the occurrence of butterflies over the period 1947–68 at Central Nyland in southern Finland and found that in years of high numbers the summer temperature was generally higher than normal but the July temperature was lower. In the preceding years the summer temperature was 0.2°C lower than normal and the July was cool. In the high frequency years the humidity was generally 14% lower than normal and in the preceding years it was 14% higher. In the low frequency years the summer temperature was lower than normal, but that of the preceding year was normal, with the July temperature especially low at 1°C below normal. The humidity was 6% higher than normal but that of the preceding summer was 16% lower than normal.

In Britain, some years have been particularly notable for large numbers of immigrant species, probably indicating favourable conditions for reproduction in the winter home of these species, with subsequent dispersal. Williams (1958) produced a table showing the incidence of reports of immigrant species from 1850 to 1955 and a number of good years stand out: 1858, 1865, 1868, 1892, 1899, 1900, 1928, 1933, 1945, 1947, 1949. 1976 was a good year for the Camberwell beauty (*Nymphalis antiopa*) but not for the more usual southern species like the clouded yellow (*Colias croceus*), indicating that the warm conditions were localised in northern latitudes, rather than over the Mediterranean (see Sec. 3.2).

Before ending this discussion of the influence of climate on population we should perhaps consider the significance of the influence of climatic fluctuations on food production in the context of the burgeoning human population. The present world population of about 4.3 billion (including an estimated 450 million people who are seriously malnourished) is currently increasing by some 70 million per year (Brown 1979) and is expected to rise to 6 or even 6.5 billion by the year 2000 (Brandt 1980) with a corresponding increase in the demand placed upon the Earth's agricultural productivity. Already total world food supplies are so precarious that starvation may result in developing countries not only from the climatically related crop failures in Africa and India that have become only too familiar in recent decades, but similarly induced harvest shortfalls in the grain-growing regions of middle latitudes (essentially in North America, Australia and New Zealand) which are the source of food imports for the rest of the world. For a consideration of this subject see Roberts and Lansford (1979) and Lamb (1977b).

2.4   Tolerance limits and changes in distribution

Each species of plant and animal has evolved particular tolerance limits to climatic conditions, as discussed in Section 2.1. These limits may be considered as defining the main parameters of the 'fundamental niche' (Hutchinson 1957) of an organism, although its 'realised niche' is constrained by biological interactions, notably competition. The tolerance limits of an organism may be determined by experimental manipulation as in the case of Crisp's (1965) work on the lower lethal limits, defined as the temperature causing a 50% mortality after a period of exposure of 18 h in the air, of a number of intertidal animals. Alternatively, the natural distributional limits of an organism may be mapped and then matched with an appropriate climatic parameter, usually a minimum or maximum isotherm. The classic exposition of this latter technique is Iversen's (1944) work on the distribution of the holly (*Ilex aquifolium*), ivy (*Hedera helix*) and mistletoe (*Viscum album*).

Holly, a species with a markedly oceanic distribution (see Fig. 2.1) is very sensitive to frost, and Iversen showed that it does not generally occur where the mean temperature of the coldest month is less than -0.5°C. Thus the species suffered a 90% mortality in Denmark during the severe winters of 1939–42, only surviving at sites near the sea where the moderating influence of the maritime climate was maximal, or in protected woodland sites. The ivy also has an oceanic distribution but is able to extend to areas where the mean temperature of the coldest month is not less than -2°C. The mistletoe can survive winter temperatures down to -8°C but demands more summer warmth than holly or ivy, requiring the mean summer temperature in the warmest month to be not less than 15.8°C as against 12.5°C and 13°C respectively for the other two species.

Oceanic species such as the three studied by Iversen are limited in their

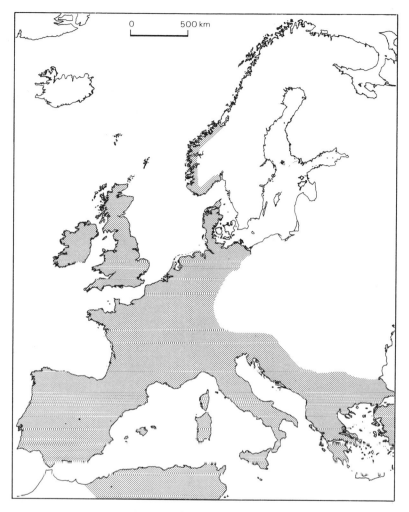

Figure 2.1 Distribution of the holly (*Ilex aquifolium*). Redrawn from Godwin (1975), after E. Dahl.

distribution by minimum winter temperatures. However, in temperate regions with a seasonal climate, many species of plants and animals die down in winter or overwinter in an inert state. Consequently, summer temperatures are more important in determining distributional limits and many European species in consequence have northern limits that are orientated in a south-west–north-east fashion, following the increase in continentality inland with distance from the moderating influence of the Atlantic Ocean, and in line with the isotherm for the mean temperature of the warmest month of the year (see Fig. 2.2). Many invertebrates, being primarily active in the summer, have a similar pattern of distribution, representing a compromise between temperature requirements and the increased moisture of more western localities.

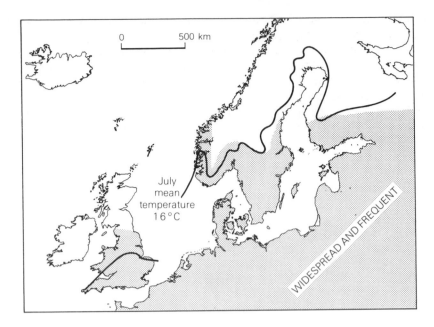

Figure 2.2 Distribution of the lime (*Tilia cordata*) in northern Europe in relation to the isotherm for mean temperature of 16°C in July, not corrected to sea level. Redrawn from Pigott (1975).

As climatic changes bring about a re-alignment of the location of the limiting isotherms (or isopleths – lines of equal rainfall – as the case may be) the distributional limits of organisms are correspondingly adjusted. Owing to their mobility, the distributional ranges of animals are much more responsive to climatic changes than are those of plants. In particular, changes in the distribution of birds (which, in addition to being highly mobile, are also visually distinctive and thus particularly well observed) have been quite well reported, the best examples being the sudden arrival of species in Iceland in response to the twentieth century climatic warming, events which can in many cases (e.g. black-headed gull (*Larus ridibundus*) in 1911, starling (*Sturnus vulgaris*) in 1941) be precisely dated.

There is often a considerable lag between the time when conditions at a previously marginal site are suitable for the survival of an organism and the actual arrival of the organism at that site. An example is the lapwing (*Vanellus vanellus*) which extended its breeding range to Iceland only in 1961 after the amelioration was essentially finished. One would expect the responses of an organism to a climatic deterioration, however, to be rather more rapid: one extreme event may be sufficient to cause a significant re-adjustment of distributional boundaries. However, unless an accurate survey is being conducted, the decline of a species may not be noticed for a considerable period of time. Bird watchers in particular are much more attentive to the arrival of a rarity than the disappearance of even a common species, so once again an apparent time-lag is manifest.

As our knowledge of the distribution of organisms in earlier decades is so meagre, it is difficult to place in their correct context some of the examples of species whose range has apparently contracted in line with the climatic deterioration during the past 30 years or so. The awakening of scientific interest in field studies and ecology essentially occurred during the period of climatic warming. For example, Charles Elton's book *Animal ecology* was first published in 1927 (Elton 1927). As such, the original data from which distributional changes have been inferred may have been atypical of the longer history of the fauna and flora and recent declines may in fact be more of a return to the *status quo ante* pertaining before the twentieth century warming. This is probably the case with the woodlark (*Lullula arborea*) in Britain, a species with a southerly distribution in the country and noted for its susceptibility to hard winters such as those of 1916/7, 1928/9 and 1946/7. Harrison (1961) compared the available data on woodlark abundance with mean annual temperature from 1880 onwards and concluded that the rise and fall of woodlark numbers is closely linked with climatic trends. In the present century the species increased in abundance and spread northwards from the 1920s and reached a peak in 1950–53 when it was breeding in forty counties of England and Wales. However, there was an almost total collapse after the 1962/3 winter and by 1968–72 breeding was confirmed in only fifteen counties (Sharrock 1974). Looking at these changes in the context of what we know about the changing climate during the present century, it is clear that the 1920–50 distribution of the species was really atypical and that the recent decline is best considered as a return to more normal conditions.

2.5 Genetic adaptation of tolerance limits

The tolerance limits of a plant or animal species, like any other genetically determined attribute, are evolutionarily adapted to the prevailing environmental conditions. If these change, natural selection will operate to bring about an alteration in the organisms' tolerance limits. By this means organisms may respond physiologically to a climatic change without any apparent change in their distribution or abundance. This kind of selection on the basis of physiological tolerance limits will proceed more rapidly in those organisms with a short lifespan and such organisms may, by this method, keep up with the changes in climatic conditions. Such a process will be difficult to detect in the wild, although laboratory experiments could usefully provide information on the rate at which tolerance limits can be selected. The fact that such selection has occurred in the past is demonstrated by the differing tolerance limits of British and continental forms of the same species. British forms generally have a narrower tolerance range than continental forms as they are adapted to a more oceanic climate with fewer extremes. Further east, the climate becomes more continental with hotter summers and colder winters and so the European forms of a given species can tolerate a greater range of climatic extremes, particularly of temperature. There is usually no physical manifestation of the differences between the two forms, which are termed physiological races, and they can freely interbreed if they come into contact with one another.

Some examples of the different tolerance limits of the European and British physiological races of certain plant species can be found in Table 13.10 ('Apparent climatic limits of the ranges of vegetation species living today') of Lamb (1977a). Thus, Lamb states that the mountain avens (*Dryas octopetala*) is limited by temperatures above 23°C in the British Isles but can withstand temperatures up to 27°C in Scandinavia. The dwarf birch (*Betula nana*) is limited to sites with temperatures no higher than 22°C in Britain but in Scandinavia is found in localities with summer temperatures up to 27°C. Mixed oak forest is dominant in areas where the mean temperature of the warmest month is not below 12–14°C in oceanic regions or 16–18°C in continental interiors. Clearly, expressing distributional relationships in terms of correlations with meteorological parameters in this fashion does not necessarily imply that the chosen parameter is the one to which the organism is responding – hence Lamb's use of the term '*apparent* climatic limits'.

A further series of examples of selection producing strains with different tolerance limits can be found amongst those exotic species introduced into temperate environments and originally found in artificially heated habitats but subsequently found in unheated conditions. The south-eastern Asian oligochaete worm *Branchiura sowerbyi* has been found in heated fish tanks in Regent's Park (London), in Kew, in Oxford and in Dublin (Macan 1974) and also in the River Thames in the warm outflow from Earley power station (Mann 1958). Aston (1968) recorded it in a warmed effluent at Coventry but also at two unheated localities: the River Otter in Devon and the Long Water at Hampton Court. Macan (1974) states that P. M. Jenkin had observed a colony for 5 years in the Kennet and Avon Canal and had recently found another in the River Avon at Bradford. Early in this century, the freshwater pulmonate snail *Physa acuta* began to spread eastwards and northwards from its original range in southern and western Europe (Frömming 1956, Macan 1974) as it was transported in consignments of tropical aquatic plants until, by 1956, Frömming stated that it could be found in every botanic garden between Amsterdam and Moscow. From these centres it has spread out into unheated waters; thus Langford (1972) records it in the Rivers Trent, Ouse and Witham in England as well as in the cooling tower ponds of a number of power stations. The trumpet ramshorn snail (*Planorbis dilatatus*) is another exotic pulmonate, this time originating in the United States. It was first recorded in Britain in 1919 in canals around Manchester where it was associated with the warm water coming from cotton mills in the area. Dance (1970) found it in Lake Trawsfynydd, Wales, which is warmed by a power station, but it has also been found in apparently unwarmed waters such as the River Tame at Dukinfield and a canal near Huddersfield (Fryer 1954).

These three species all originated in, and were presumably adapted to, waters with temperatures considerably above those prevailing in natural freshwater habitats in Britain to which they ultimately adapted. For example, Aston (1968) found that immature specimens of *Branchiura sowerbyi* grew fastest at temperatures between 25 and 30°C. On being introduced to Britain, they were first reported from artificially warmed waters, such as power station effluents, although subsequently they were recorded in apparently unheated waters. This suggests that selection had acted to reduce the thermal tolerance limits of these species to those temperatures generally prevailing in natural freshwater habitats in Britain. These three examples all concern species found in freshwater habitats where temperature is the most important climatic factor. Examples from terrestrial systems are not well recorded, although the common or oriental cockroach (*Blatta orientalis*) should be mentioned. This exotic species was probably established in houses in England by the fifteenth century (Marshall 1974) and formerly it was thought that cockroaches could not survive outdoors unless in some artificially heated habitat such as a compost heap. However, more recently colonies have been reported breeding in a number of unheated situations (Beatson & Dripps 1972), suggesting that the insect's tolerance limits have become adapted to lower temperatures.

3

Recent climatic changes and
their possible causes

3.1 The nature of twentieth century climatic changes

The principal feature of our changing climate over the past 100 years is that of
rising temperatures reaching a peak in the first half of the twentieth century
which had not been attained since mediaeval times. In fact temperatures had
been gradually rising since the eighteenth century (see Sec. 4.5) but the trend
did not become manifest until the 1850s when recession of the Alpine glaciers
began. There was a slight setback from 1875 to 1890 when there was a return to
cooler conditions with Dickensian winters due to stationary high pressure
features over Scandinavia directing cold polar air southwards.

During the early part of the twentieth century there was a general rise in
mean annual temperature averaged over the whole northern hemisphere of
about 0.5°C (see Fig. 3.1). This was initially effected by an amelioration of

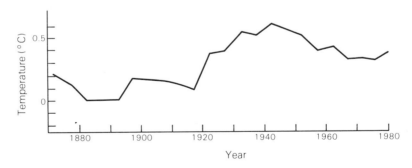

Figure 3.1 Mean annual surface temperatures of the northern hemisphere: estimated
changes in 5 year averages over latitudes 0–80° N since 1860. Redrawn from Lamb
(in press).

winter temperatures with, for example, a virtually unbroken run of mild
winters in the British Isles from 1896 to 1939. Later, other seasons were
affected culminating in central Europe in the long, warm summers of the two
decades between 1933 and 1952. The twentieth century warmth in northern
Greenland peaked in about 1930, rather earlier than in Europe, as was the case
in the earlier mediaeval warm period referred to above. As well as being

registered earlier in higher latitudes, the magnitude of the twentieth century climatic changes were greater further north. Thus the temperature changes for each 5 year period from 1870–74 to 1955–59 averaged over the latitudinal zone 60–80°N appear to be about four times as big as the average for the whole Earth and about ten times as big as for the equatorial zones (Mitchell 1961). Owing to the position of the British Isles in the middle latitudes on the western edge of a continental mass, the changes affecting the averages over a decade or longer at this location for this and earlier centuries are probably representative of the averages for the whole northern hemisphere and even the whole Earth: the temperature trend from about 1880 to the present has been generally similar in both hemispheres down to about 35–40° latitude, although it seems to have been in antiphase nearer the poles.

Since about 1940–50 there has been a lowering of overall northern hemisphere average temperatures by about 0.3°C (a trend equivalent to that between the 1730s and the 1760s). The mean temperature in England had fallen by this amount by the 1960s but here, as in much of Europe and eastern North America, recovered for a few years between 1972 and 1976: again there seems to be some parallel with the 1770s. The reversal of the twentieth century warming was first seen in the decline of winter temperatures associated with the establishment of a high pressure ridge over Greenland (Dickson, Lamb, Malmberg & Colebrook 1975) which directed a flow of cold polar air southwards over Britain and Europe. The Arctic drift ice became more extensive and in April and May 1968 it advanced around Iceland as it had not done before since the early years of the century. The northern coast of Iceland is usually kept free of ice by the influence of the warm Irminger Current (a distant extension of the Gulf Stream), but in this and subsequent years the East Greenland Current flowing from polar waters predominated and carried the drift ice with it. This invasion of cold water caused the herring (*Clupea harengus*) to leave the area and migrate eastward, and the consequent drop in Iceland's herring exports was held to have caused two devaluations of the country's currency (Kristjansson 1969). Satellite photographs of the northern hemisphere (which began in 1967) showed a further, and more widespread, increase in the drift ice in 1971, during the course of which year annual mean snow and ice coverage increased by 12% (Kukla & Kukla 1974) and subsequently remained high. In addition to sudden incursions of cold water, the general trend of sea temperatures has broadly followed the changes in air temperature during the present century, although usually with a lag of 10–20 years associated with the high specific heat of water (i.e. the large amount of heat required to raise the temperature of a given mass of water) together with the continual stirring of the top 10–100 m of the ocean by wave action.

3.2   The atmospheric circulation and twentieth century climatic changes

In considering the nature of twentieth century climatic changes we have so far concentrated on temperature changes. This is a reflection of the fact that the global climatic system is driven by the temperature difference between the equator and the poles: the unequal heating of the atmosphere by latitudes, resulting from the spherical shape of the Earth, causes an expansion of the air at the equator producing at that latitude a zone of high pressure in the middle and higher levels of the atmosphere, and areas of low pressure at the poles. This pressure difference tends to produce a high altitude movement of the air from the equator to the poles. However, the Earth, and the atmosphere surrounding it, is rotating eastwards once every 24 h. The speed of rotation of the atmosphere is greatest (about 25 000 miles or 40 000 km per day) at the equator and zero at the poles. A mass of air moving from the equator towards the poles retains its original eastwards momentum relative to the Earth's surface so that the initial Equator-to-pole movement is translated into a west-to-east movement in both hemispheres. In consequence, the upper winds flow *along* rather than *across* the lines of equal pressure. This high altitude (operating between 2–3 km and 15–20 km) flow of westerly winds (i.e. moving eastwards – winds are designated relative to their source rather than the direction towards which they move, in contrast to ocean currents) is the main feature of the atmospheric circulation in both hemispheres, controlling the development of, and the tracks followed by, surface weather systems. It is strongest over middle latitudes where it is termed the jetstream or circumpolar vortex. In this region (latitudes 40–60°C) the prevailing *surface* winds as well as the upper winds tend to be westerly. The circumpolar vortex generates high pressure systems (such as the Azores anticyclone) at the surface on its warm, equatorial side and low pressure systems (such as the Iceland low) at the surface on its cold, polar side.

The upper westerly flow develops a number of waves or meanders which are especially marked in the northern hemisphere. These are called ridges when they veer polewards, carrying warm tropical air from lower latitudes, and troughs when they bulge equatorwards, bringing cold polar air from higher latitudes. The number of waves (called Rossby waves after the Swedish meteorologist of that name – see Rossby 1939, 1941) can vary from two to about six and their amplitude can also vary. The system tends to be anchored by the high mountains of the Rockies which extend into the upper atmosphere and induce a ridge in the upper westerly flow by causing the jetstream to veer northwards. The appropriate number of waves (determined by the wavelength favoured by the speed and latitude of the flow) are then fitted in downstream. At times when the equator–pole temperature difference is greatest (e.g. in winter) the upper westerly circulation is generally strongest, with few meanders, and the number of waves is low (i.e. the wavelength is long). When the north–south temperature gradient is reduced (as in the northern summer) the atmospheric circulation is less vigorous and there are more waves (of shorter wavelength) in the upper westerlies.

Direct observation permitting daily surveys of the upper winds has only been possible for the last 30–40 years, but a useful analogue of the strength of the upper westerlies is provided by the index of the number of days per year of westerly weather experienced in the British Isles whose geographical position in middle latitudes at the eastern limit of a large flat (i.e. no vertical geography to deflect the winds) ocean means that surface winds generally register what is going on in the upper atmosphere. The westerly weather type, according to Lamb's classification of circulation patterns (Lamb 1972a), corresponds with south-westerly surface winds and changeable weather, mainly cool in summer and mild in winter, and an annual index of the number of these westerly days, such as is presented in Figure 3.2, is a useful indicator of the vigour of the global circulation. A rather similar index going further back in time is shown in Figure 3.3, which represents the frequency of south-westerly surface winds

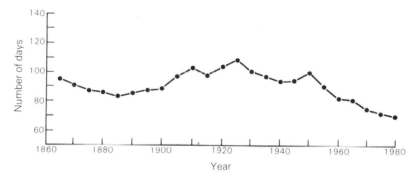

Figure 3.2   Annual index of westerly days over the British Isles: 10 year average plotted at 5 year intervals. Redrawn from Lamb (In press).

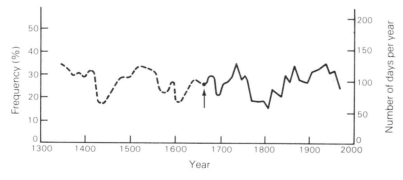

Figure 3.3   Frequency of south-westerly surface winds in south-eastern England since 1340. Ten year averages compiled from daily observations at London since 1670; estimates from indirect evidence back to 1340. Redrawn from Lamb (1967).

in south-eastern England since AD 1340. This is compiled from a series of
overlapping daily records of the winds in London since 1670 and from earlier
indirect evidence from a variety of weather diaries going back to 1340.

Examination of Figure 3.2 indicates that the atmospheric circulation was
more vigorous during the earlier decades of this century and then declined in
strength. This would, at first sight, suggest that there was an increase in the
equator–pole temperature gradient which would have been registered as a
high latitude cooling. From the account of twentieth century temperature
trends presented in Section 3.1 it will be recalled that this was not in fact the
case. On the contrary, during this period of strong circulation the gradient of
surface temperatures decreased owing to the warming of the Arctic region.
However, there is evidence (see Lamb 1972b, pp. 95–6) that the gradient of
upper air temperatures between the equator and the pole was in fact greater
during this period: upper air temperatures rose proportionately more in lower
latitudes than at the pole, possibly indicating some change in the output of the
Sun or in the amount of radiation reaching the troposphere (the lower
atmosphere, which is separated by the tropopause from the upper atmosphere
or stratosphere) by a decrease in the amount of stratospheric volcanic dust
(see Sec. 3.4). Similarly, during the period 1945–65 when the atmospheric
circulation was weakening, the cooling of the upper levels of the air was most
pronounced in the subtropical zone (Lamb 1972b). Thus it is best to consider
the climatic changes of the first half of the twentieth century as being driven by
an increase in the Equator–pole temperature gradient of the upper
atmosphere which produced an increase in the vigour of the atmospheric
circulation which then spread the warming of surface temperatures to the
highest latitudes.

Concentrating on the latitude of the British Isles, the westerly winds that
predominated during this period had travelled across the Atlantic Ocean
which, owing to the thermal properties of water, exerts a moderating influence
on the temperature of the air passing over it. The ocean reduces the higher air
temperatures in the summer and raises the lower air temperatures in the
winter. In consequence, the climate of Britain and north-western Europe was
notably equable, without extremes of temperature. The westerly winds also
absorbed moisture as they travelled across the Atlantic and so rainfall was
more abundant and consistent. Rainfall penetrated further into most
continental interiors, apart from the American Mid-West where vigorous
westerlies intensify the rain-shadow effect of the Rockies, and the monsoons
of India and West Africa were more reliable. The increased vigour of the
atmospheric circulation also affected the strength of the ocean currents which
are mainly driven by the surface winds.

Since about 1940, when temperatures started to drop, the westerly
circulation has declined in vigour and has become more erratic and often more
meridional (north–south in character as distinct from east–west or zonal) with
more pronounced ridges and troughs. With this sluggish type of circulation a
high pressure anticyclone may be cut off at the poleward tip of a warm ridge or

Table 3.1   Some notable extremes of climate since 1960 (after Lamb 1977b).

Year	Climatic extreme
1961	Extremely high equatorial rainfall in East Africa: East African great lakes rose in a few months to above all twentieth century records.
1962–3	Coldest winter in England since 1740.
1962–5	Driest 4-year period in the eastern United States since records began in 1738.
1963–4	Driest winter in England and Wales since 1743; coldest winter over an area from the lower Volga basin to the Persian Gulf since 1745.
1964	Snow covered all the uplands of South Africa and Namibia (South West Africa) in June, the heaviest and most widespread snowfall there since 1895, causing many deaths.
1965–6	Baltic Sea completely ice-covered.
1966	Arctic pack-ice reached the supposedly ice-free port of Murmansk, on the south coast of the Barents Sea, for the first time known.
1968	Ice half-surrounded Iceland and stopped shipping for the first time since 1888.
1968–73	Severest phase of the drought in Ethiopia and the Sahel region of Africa.
1960–9	Driest decade in central Chile since the 1790s.
1969	Lowest frequency of westerly wind days (56) in British Isles for at least 109 years, and possibly since 1785.
1971	Barometric pressure map for September showed anomalies in three areas, North America, North Atlantic and Siberia, amounting to five standard deviations from the average values for the earlier part of the century.
1971–2	Coldest winter on record in parts of eastern European Russia and Turkey: River Tigris frozen over in eastern Turkey.
1972	Highest temperatures in summer ever recorded in northern European Russia and Finland (33°C in Lapland); great drought caused major shortfall in Soviet grain harvest. Number of icebergs (1587) on the western Atlantic south of 48°N exceeded that in any previous year since records began in 1880.
1973	Great Lakes of North America and the Mississippi River at highest level since 1844. Severest drought in central America for many years. First ever report of snow on high ground in Queensland, Australia.
1973–4	In central Australia unprecedented floods in January ended succession of great drought summers.
1974–5	Mildest winter in England since 1834; least ice on Baltic Sea since perhaps 1652.
1975	Snow in London in June, followed by great heat wave in July and August in Britain and western Europe, especially 4–11 August when mean temperatures for the week in the Netherlands and Denmark (approx. 24°C) exceeded previous highest by over 2°C. Arctic sea ice returned to Iceland in July for the first time in the present century.
1975–6	Great drought in northern Europe (water transport on the Rhine reduced as the river level dropped) and especially in England where rainfall from May 1975 to August 1976 was lowest 16 month total since records began in 1727.

1976	Great heat in June–early July in western Europe: temperatures over a 24 day period in England exceeded the highest monthly mean in the 300-year record by about 4°C. Very cold, wet summer in European Russia and parts of Canada.
1976–7	Record cold temperatures and heavy snow in eastern and mid-western United States; all time recorded low of -3°C at Palm Beach; snow in Florida.
1979	Severe drought in north-eastern Brazil, continuing into 1980.
1980	Hottest and driest summer in memory in south-eastern and mid-western United States, more than 1000 deaths from heat in late June and July.

a low pressure cyclone may be separated out from the southernmost extent of a cold trough. These stationary weather features appear to block the flow of the westerlies and result in long-lasting runs of the same kind of weather, respectively hot or dry and cold or wet. The effect of such blocking is to produce climatic extremes of one kind or another, such as those listed in Table 3.1. Such blocks will occur at different longitudes at different times, depending on the predominant wavelength in the circumpolar vortex, and this will result in an increase in the variability of weather experienced at a particular site. Wallén (1953) examined temperature records for Sweden over the past 200 years and found that the variance of summer temperature was greatest in periods of meridional circulation. Similarly Lamb (1977a, p. 486) has shown that the variance of winter temperatures in central England tends to be greatest in periods when westerly winds are least prevalent. Ratcliffe, Weller and Collison (1978), however, examined the variability of three parameters of the weather over Britain (surface pressure anomalies, surface temperatures of central England and rainfall for England and Wales) over about the last 100 years and concluded that there was very little evidence for any unusual variability in climate in recent years on annual, monthly or pentad (5 day) time-scales, although their analysis clearly indicated the extremely anomalous nature of certain years such as 1975 and 1976. This work also produced evidence of a circulation change around 1940 in the eastern Atlantic–British Isles sector which, as described above, can be considered the prime indicator of the changing climate.

As well as temperature variability, rainfall variability also increases as the westerlies decline: in 1972 a blocking anticyclone over eastern European Russia produced a serious drought and disastrous grain harvest in that area; in 1975 and 1976 (when European Russia experienced a cold, wet summer) the block was located further west, centred near southern Britain, producing a notable drought (Doornkamp 1980) as the rain-bearing depressions were deflected northwards towards Iceland and the extreme north-west of Scotland. The period May 1975 to August 1976 was the driest 16 month period in Britain since rainfall records began in 1726 but it was immediately followed by the wettest autumn on record and the fourth coldest December, emphasising the variability inherent in the current climatic situation. Similarly, 1976 was the second warmest summer in the long temperature record for central England going back to 1659 (see Sec. 4.5) whilst the following year, 1977, equalled the eighteenth coldest. The drought summer of 1976 exemplifies the dangers of drawing general conclusions about climate from one's own immediate experience: as Figure 3.4 shows the area of reduced rainfall was really quite localised and over the hemisphere as a whole the areas in which the 1976 summer was colder than the 1931–60 average exceeded those where it was warmer (Fig. 3.5). Similarly, although Britain and much of Europe experienced a run of mild winters in the early 1970s, temperatures were persistently below average in North America and for the hemisphere as a whole the cooling trend was maintained. This run of mild winters was

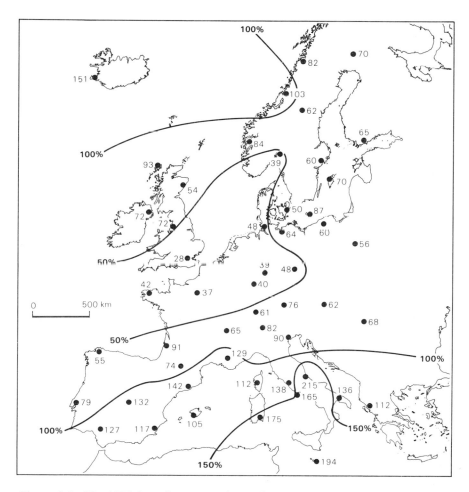

Figure 3.4 The 1976 drought summer in north-western Europe: February–August rainfall expressed as a percentage of the 1931–60 average. Redrawn from Jones (1976).

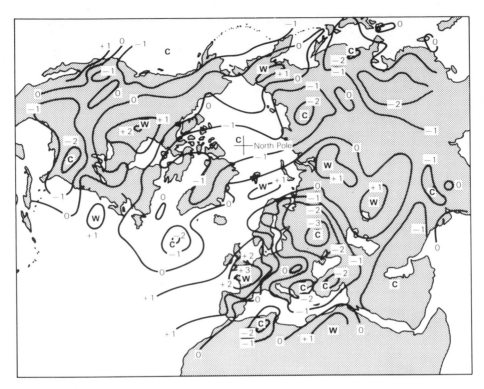

Figure 3.5 Northern hemisphere temperatures for summer 1976: June, July and August mean temperatures expressed as the departure, in degrees Celsius, from the 1931–60 average. Redrawn from Lamb (in press).

associated with the collapse of the high pressure ridge over Greenland which had produced a northerly airflow in winter during the 1950s and 1960s (Dickson *et al.* 1975) and the transfer of the hemispheric centre of cooling from the Siberian to the Canadian sector where there was renewed growth of snowbeds and glaciers. The difference between the winters of the early 1970s and those of the 1960s reflects a change in the position of the waves in the upper westerlies and their associated ridges and troughs.

3.3 Recent trends in other parts of the world

The central role of the upper westerly winds in driving the atmospheric circulation and linking the climatic phenomena experienced at different longitudes tends to focus attention on events in middle latitudes. However, the tendency for the westerlies to decline in vigour and for the circumpolar vortex to become more meridional has had worldwide effects, particularly on the distribution of rainfall which has not penetrated so far into the continental interiors, with resultant failures of the monsoons of India and Africa. Interestingly enough, it was the perceived decline in the population of a European bird species which first alerted scientists to a possible mechanism by which changes in the circumpolar westerlies may bring about changes in the rainfall in the tropics. The bird concerned is a small warbler, the whitethroat (*Sylvia communis*), which visits Britain and north-western Europe between April and September. It is one of the more common warbler species in Britain (together with the willow warbler, *Phylloscopus trochilus*) and inhabits a variety of scrubby situations including hedges, heathland and woodland edges. Although populations generally fluctuate from year to year, numbers in Britain (as measured by the Common Bird Census described in Sec. 2.3) decreased dramatically by 77% in 1969, with numbers remaining low in subsequent years (see Fig. 3.6a). Similar declines were noted over continental Europe (Berthold 1973). In autumn the whitethroat migrates to the Sahel zone of West Africa, that is the area south of the Sahara between 10°N and about 20°N. The climate of this region is characterised by a summer monsoon produced by the south-westerly winds blowing inland from the Atlantic Ocean. A monsoon is a large scale, seasonally changing wind phenomenon analogous to the daytime sea breezes experienced along coasts: the land heats up more rapidly than the sea and the air above it rises, drawing in moist air from the sea at the surface. In the case of a monsoon, as the moist air rises over the warmer land, its moisture condenses out as rain along the Inter-Tropical Convergence Zone (ITCZ). In winter, the increased vigour and lower latitude of the strongest upper westerly circulation means that the troughs penetrate further into subtropical latitudes and so force the equatorial rain belt that produces the monsoon away to the south. As the rain-bearing ITCZ migrates northwards and back again between May and October, there is a marked north south gradient in mean summer rainfall from 50 mm near the Sahara to 1500 mm near 10°N (see Fig. 3.6b), giving a gradient of 1 mm km^{-1} (World Meteorological Association 1976). Furthermore, the rainy season begins later and ends earlier the further north one progresses. Since 1960 the ITCZ has not moved as far north and so the zonally arranged isohyets (lines connecting places with equal amounts of rainfall) have been gradually moving south at a rate of about 9 km a^{-1} with the result that total May–October rainfall over the area has been declining (see Fig. 3.6c). This results in a reduction in the amount of plant and animal food available to the whitethroat and other migratory species such as the sand martin (*Riparia riparia*) and sedge warbler (*Acrocephalus schoenobaenus*); a reduction which is most critical at the end of the dry season when the birds need to build up their energy reserves for their return flight to Europe. In 1968, the year before the decline of the whitethroat was first noticed in Europe, precipitation in the Sahel zone was, on average, 25% below normal and in places 70% below normal. In 1969 average rainfall was 10% below normal and whitethroat numbers in Britain picked up slightly (Fig. 3.6a). However, in 1970 rainfall was 23% below normal, in 1971 36% below normal and in 1972 39% (Winstanley *et al.* 1974).

The decline of the whitethroat noticed in Europe was an early indicator of
the major drought in the Sahel region (primarily the countries of Mauritania,
Senegal, Mali, Upper Volta, Niger and Chad, although the drought extended
right across Africa to Ethiopia) which between 1968 and 1973 led to the deaths
of 100 000–250 000 people (United Nations 1977). Winstanley (1973a, b) has
suggested'how the reduced rainfall in this area can be interpreted in terms of
the recognised decline of the atmospheric circulation in higher latitudes:
according to his hypothesis, the abnormal southward extent of the troughs in
the meandering circumpolar westerlies has restricted the northward extent of
the tropical meridional circulation systems controlling the monsoon rainfall

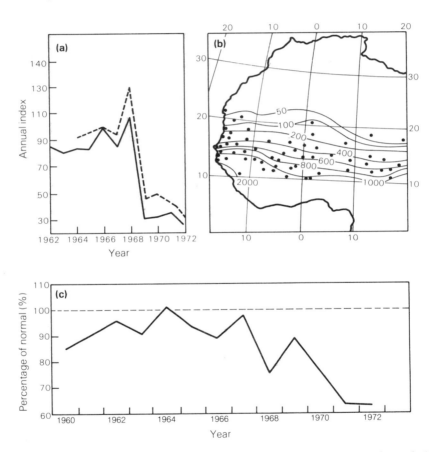

Figure 3.6 The effect of the Sahel drought on the British population of the
whitethroat (*Sylvia communis*). (a) The annual index of whitethroat abundance
as measured by the Common Bird Census: — farmland plots; ---- woodland plots.
(b) Average May–October rainfall isohyets for 1931–60 (in millimetres) for the
Sahel zone of Africa; dots indicate locations of rainfall stations. (c) Mean percentage
of 1931–60 May–October rainfall for the sixty stations in the Sahel zone of Africa
represented by dots in (b). Redrawn from Winstanley, Spencer and Williamson (1974).

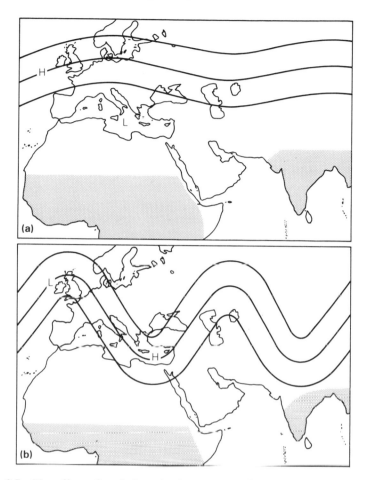

Figure 3.7 The effect of variations in the strength of the atmospheric circulation in higher latitudes on the extent of the monsoon rainfall (stippled) of the Tropics. (a) Strong westerly (zonal) circulation in middle latitudes permits monsoon rains to extend well northwards in Africa and India. (b) Weak westerly (meridional) circulation in middle latitudes results in southward extension of troughs and restriction of northerly extent of monsoon rains. H = high rainfall; L = low rainfall. Redrawn from Winstanley (1973b).

in West Africa as indicated in Figure 3.7. A similar effect downstream at the next trough would affect the monsoon of north-western India. Winstanley was able to corroborate his hypothesis by demonstrating that the fluctuations in the number of westerly days in Britain as recorded in the Lamb index which, as reported above, is an indicator of the vigour of the circumpolar westerlies, are largely paralleled by fluctuations in the monsoon rainfall experienced in both the Sahel and north-western India. The period of relatively high rainfall in the Sahel was from the end of the first decade of the century to the mid-

1950s, with the peaks coinciding closely with those of the frequency of the westerly weather type over the British Isles.

Although it is often stated that the Sahel drought lasted from 1968 to 1973, rainfall has been below normal again in most recent years and the drought appears to have spread to equatorial Africa as well, where its effects, combined with those of political instability, have caused widespread famine in areas such as the Karamoja region of Uganda.

3.4   Mechanisms of climatic change

Having considered the climatic changes that have been manifested during the present century, it is appropriate to review the possible mechanisms of these and similar changes which can operate over a number of different time-scales.

The theory of plate tectonics which describes the Earth's crust in terms of rigid plates which are moved apart by the spreading of the ocean floor and which provides a sound basis for Wegener's (1912) ideas of continental drift has profoundly influenced the geological sciences over recent decades. The theory has an obvious climatic dimension as the latitudinal position of a given continent will clearly have a considerable effect on the climate experienced by that continent. However, the disposition of the land masses may also exert an effect upon the climatic regime experienced by the Earth as a whole, mainly as a result of changes in the planet's reflectivity or albedo. It is generally accepted that when land masses become concentrated at high latitudes they tend to become coated with a layer of snow and ice which reflects back more of the incident solar radiation with the result that global temperatures are lowered by the reinforcing effect of positive feedback (although see Cogley 1979, who suggests that, as the land–water albedo contrast is greater in the tropics, the Earth cools when land masses are concentrated at low latitudes). Computer simulations of an equatorial continental belt on the one hand, and the land masses concentrated into polar caps on the other, produces a 12°C difference in mean surface temperature (the latter configuration being the colder), while there is probably only a 6°C difference (on a global average) between recent glacial and interglacial regimes (Sellers & Meadows 1975). Mechanisms of climatic change involving positive feedback have an attractive logicality until one realises that they appear to offer no means of getting out of an ice age of ever increasing severity. At this point one can then invoke the influence of *negative* feedback: as the oceans get cooler, less evaporation occurs, which results in less precipitation of snow and eventually winter snow accumulation is less than summer melting and the climate gradually warms.

Another mechanism of climatic change which operates over the longer term is the variation in the Earth's orbit, the shorter term manifestation of which accounts for the annual progression of the seasons. As the axis of the Earth is tilted at about 23.5° to the vertical, alternately one hemisphere and then the other is tilted towards the sun, resulting in the seasonal cycle of the year. The eccentricity of the Earth's orbit also means that the distance of the Earth from the Sun varies from 91 million miles (146.5 million kilometers) at perihelion (point of closest proximity to the Sun) on 3 January to 94 million miles (151 million kilometres) at aphelion (point farthest from the Sun) on 4 July. The Earth as a whole receives 7% more radiation at perihelion, thus moderating the cold of the northern hemisphere winter in comparison with that of the southern hemisphere. In addition, however, there are long-term cycles in the Earth's orbit. The ellipticity of the orbit varies cyclically over 90–100 ka, the tilt of the polar axis varies from 21.8° to 24.4° from the vertical and back again over 40 ka (the greater the tilt, the greater the seasonal amplitude of temperature), and there is a third cycle describing the orbital position of the Earth at the solstices and equinoxes which is termed the precession of the equinoxes and has a periodicity of about 20 ka. The Yugoslavian climatologist Milutin Milankovitch (Milankovitch 1930, see also Vernekar 1968) calculated the implications of these cycles in terms of the incident solar radiation at different latitudes and they appear to be able to account for the onset of the ice ages and the various warm peaks over the past 1 Ma. It is thought that the initially small changes in mean surface temperature resulting from the variation in solar energy input increase the snow and ice cover at higher latitudes. This increases the total albedo of the Earth and so more solar radiation is reflected back, producing a positive feedback as discussed above. Milankovitch's theory has been validated by the work of Hays, Imbrie and Shackleton (1976), who found distinctive peaks at about 100, 42 and 23 ka in the temperature profile deduced from the isotopic composition of the calcareous tests of small planktonic protozoans called foraminifera (see Sec. 4.2 for an explanation of this technique) going back over the past 450 ka obtained from deep sea sediment cores. Forward extrapolation of Milankovitch's theory indicates that snow and ice cover will reach their maximal extent again (i.e the peak of the next ice age) in about AD 25 000 (Imbrie & Imbrie 1980).

Milankovitch's theory attributes climatic changes to variations in the input of solar radiation to the Earth as a result of variations in the Earth's orbit. The possibility of variations in the *output* of radiation from the Sun has not been seriously considered until recently, hence the term 'solar constant' applied to the intensity of the solar beam at the Earth's distance from the Sun (equivalent to 1353 W m^{-2} or 1.94 cal cm^{-2} min^{-1}). However, Pivarova (1968) has reported a decline in the intensity of the solar beam of at least 4% since 1945, which might be related to recent climatic trends. There are also regular variations in the 'Sun tides' raised on the Sun by the gravitational effect of the planets, occurring with periodicities of 12, 24, 850 and 1700 years (Brown 1974). It is believed that planetary effects such as these may influence the Sun's output of radiant energy as they are associated with the development of sunspots. Sunspots are areas on the surface of the Sun which are rather cooler than the rest of the photosphere. The emission of radiant energy is reduced by 50–75% in the vicinity of a sunspot, but an increase in sunspot activity is generally accompanied by an increase in solar faculae and flares, suggesting that high sunspot activity is an index of an increased flux of radiant energy from the Sun, probably associated with increased convective transport of hot gases from inside the Sun to the surface of the photosphere. It is perhaps significant that the 'Maunder Minimum', a period from 1645 to 1715 when there were apparently no sunspots visible from the Earth, corresponds with the height of a particularly cold episode in the Earth's Postglacial climatic history which is known as the Little Ice Age (see Sec. 4.5). The occurrence of sunspots varies with an irregular period of about 11 years and there is also a longer periodicity of about 180 years in the magnitude of the maxima (e.g. the very large maxima of 1778 and 1957) and the length of the cycles – the strongest activity cycles may be as short as 9 years. The relationship between sunspot numbers and climate is rather complex and sometimes correlations are only valid over relatively short periods of time or for particular parts of the globe. For example, the level of Lake Victoria in Uganda (reflecting the seasonal rainfall over East Africa) showed a high correlation with the sunspot index from 1890 to 1930, at which point the relationship changed to a double oscillation with each solar cycle (Lamb 1966). Similarly, although wheat production over much of the northern hemisphere is enhanced around sunspot maxima (a relationship probably correlated with rainfall as well as temperature as the droughts in the mid-western United States seem to be associated with every second sunspot minimum), the opposite relationship holds for the southern hemisphere.

Having dealt with variations in the input of solar energy to the upper atmosphere, the next variable that should be considered is the transmission of this energy to lower levels. Transmission varies on a daily basis according to cloudiness, but volcanic dust veils associated with eruptions, which sometimes persist for several years in the stratosphere, may also cause longer-term reductions in energy transfer (Bryson & Goodman 1980; Lamb 1970, 1971). One of the first people to note the relationship between an input of volcanic dust to the atmosphere and a subsequent deterioration of the weather was Benjamin Franklin who, during his stay in Paris as first US envoy, attributed the severe winter of 1783/4 to the persistent fog over Europe and the greater part of North America resulting from the major 1783 eruption of Laki in Iceland. In general the climatic effects of volcanic eruptions tend not to be very persistent, e.g. the effects of the eruption of Tambora in Indonesia in 1815 were only really manifest in the following year, 1816, which was termed 'the year without a summer' by contemporary writers. Even the effects of the major 1883 eruption of Krakatoa only persisted for about three subsequent years, this being the residence time of the ejected particles (estimated to have reached a height of 80 km) in the atmosphere. Similarly, although the eruption of Mt St Helens volcano in Washington State in May 1980 was one of the most powerful this century, its effects on world temperatures are unlikely to be persistent. However, over longer time-scales, variation in the incidence of waves of volcanic activity may produce trends in the changing climate. Thus the increased atmospheric.transparency associated with the lack of major volcanic eruptions between that of Katmai, Alaska, in 1912 and that of Mt Agung, Bali, in 1963 has been invoked as a causative factor in the twentieth century climatic amelioration. Looking further back in climatic history, Kennett and Thunnell (1975) examined the volcanic ash of past eruptions preserved in stratigraphic sequence in deep sea sediments and concluded that there has been a global increase in volcanic activity in the last 2 Ma which may be related to the frequent glaciations during this period. However, it may be that climatic changes bring about an increase in volcanic activity rather than vice versa, by increasing the strain on the Earth's crust as a result of the unloading of the oceanic crust and the increased loading of the continental crust as sea levels drop and ice sheets build up. Evidence for this sequence of events is provided by the demonstration (Ruddiman & Glover 1974) that in the northern Atlantic sector there was an increase in volcanic activity between 9 and 11 ka BP (before present), i.e. about the *end* of the last glaciation. This is attributable to the changes in stress on the Earth's crust brought about by the rapid melting of the ice sheets.

In addition to the natural emission of volcanic dust, the activities of Man also contribute to the input of particulate matter into the atmosphere: the so-called anthropogenic dust. The American climatologist Reid Bryson, in particular, has attributed the recent global cooling to a significant rise in atmospheric dust concentrations due to human activities such as the burning of fuel, slash-and-burn agriculture and the over-usage of both arable and grazing land leading to soil deflation (Wendland & Bryson 1970). Other scientists, however, consider that on balance the activities of man are tending to produce a warming trend, notably by means of the 'greenhouse effect' resulting from the increased output of carbon dioxide CO_2 by the burning of oil, coal and other fossil fuels. The CO_2 in the atmosphere permits the transmission of the incoming solar radiation but absorbs the long wavelength infra-red component of the radiation which is reflected back from the surface of the Earth. This leads to a warming of the air near the Earth's surface – the greenhouse effect. The CO_2 concentration of the air is presently about 335 ppm (parts per million – by volume), having increased from about 280 ppm at the end of the last century. Calculations indicate that if all the CO_2 released from the burning of fossil fuels over this intervening period had remained in the atmosphere, the increase in concentration would have been about twice that recorded; the 'missing' 50% having been dissolved, mainly in the form of bicarbonate ions, in the oceans, which together with the biosphere, where considerable amounts of CO_2 are locked up in the form of organic molecules by the process of photosynthesis, acts as a CO_2 'sink'. Extrapolation of the annual atmospheric CO_2 increments recorded since 1958 on the summit of Mauna Loa, Hawaii (Machta 1972), suggests that atmospheric levels of CO_2 will reach 380 ppm by the year 2000, which might produce an increase in the temperature of the lower atmosphere of 0.5°C (Manabe & Wetherald 1975). Computer simulation models indicate that a doubling of the atmospheric CO_2

concentration (which on a projection of present trends would occur in about the year 2050) would probably result in an increase in mean global temperature of some 2–3°C, which represents a climatic change greater than any experienced over the last 10 ka. It is estimated (Manabe & Wetherald 1975) that the temperature increase at high latitudes would be about three times the global average, with positive feedback of melting snow and ice cover.

Another method of obtaining an insight into the effects of possible future climatic changes, alternative to that of computer modelling, is the use of empirical climatic data for past periods as analogues of future trends. Thus, in order to examine geographical patterns of temperature and precipitation changes which may follow a global warming such as it is thought would follow a large increase in atmospheric CO_2 levels, Wigley, Jones and Kelly (1980) compared the pattern of temperature and precipitation changes between a composite of the five warmest years (1937, 1938, 1943, 1944 and 1953) in the period 1925–74 and a composite of the five coldest years (1964, 1965, 1966, 1968 and 1972) in the same period. The average temperature difference over the northern hemisphere between the warm and cold year groups was found to be 0.6°C, but for high latitudes (65–80°N) it was 1.6°C. Considering the pattern of the composite warm period as an analogue for the circulation features likely to be associated with a global warming, it is interesting that although temperature increases are indicated for most regions (especially in high latitudes and the continental interiors such as much of North America), some regions (much of India, Iberia and northern Africa) show temperature decreases. With respect to precipitation changes, which are always much less spatially coherent than temperature changes, although there was a slight overall increase in rainfall associated with the warm year group, with a major increase over India related to a more intense monsoon circulation, decreases indicated over much of the United States, Europe and Russia might have major implications for agricultural productivity.

It seems likely that the CO_2 concentration of the atmosphere has undergone considerable natural variations in the past: analysis of air bubbles trapped in polar ice (Delmas, Ascencio & Legrand 1980) indicates that at the height of the last ice age (about 18 ka ago), the atmospheric CO_2 content was about half the present value. This demonstrates the difficulty of establishing cause and effect as it is unlikely that the lower temperatures of the ice age were caused by a reduced greenhouse effect consequent on lowered CO_2 concentrations (brought about by what mechanism?). Rather, it is more probable that the reduced CO_2 concentration in the atmosphere resulted from some effect of the onset of the glaciation (perhaps associated with an increase in photosynthesis activity in the less extensive oceans).

Although considerable attention has been paid to the probable effect of CO_2 concentrations in producing significant climatic changes in the near future, the effects of an increasing level of atmospheric CO_2 may not be so deleterious as has sometimes been suggested. For example any warming produced by the man-induced increase in CO_2 may be cancelled out by the *natural* cooling of the world climate which many climatologists consider to be the probable trend for the immediate future. Furthermore, any initial temperature rise would be offset to some extent by the shading effect of increased cloudiness consequent on the enhanced evaporation from the oceans. In addition, although it is suggested that the distribution of natural vegetation will change in response to the CO_2-induced climatic changes and that agriculture will be disrupted, there is considerable experimental evidence that photosynthesis is in fact limited by low levels of ambient CO_2. An increase in CO_2 concentration may well therefore produce an increase in photosynthetic rate and hence result in a net increase in the agricultural productivity of the world. Finally, although the greenhouse effect produces a warming of the atmosphere near the Earth's surface, the temperature of the stratosphere will drop as less infra-red energy reaches the higher levels. The resulting temperature decrease will produce a change in the rates of some of the chemical reactions controlling the production of ozone high in the atmosphere and will lead to an increase in the total ozone content of the stratosphere (Groves, Mattingley & Tuck 1978). This will help offset any depletion of the ozone caused by chemical reactions induced in the stratosphere by the terrestrial release of chlorofluorocarbons (CFCs) (which may in addition have a direct influence on the radiation balance of the atmosphere) from aerosol cans and refrigerators (Isaksen, Hesstvedt & Stordal 1980). The stratospheric ozone layer generally screens out carcinogenic ultra-violet rays of wavelengths 290–320 nm and so the major effect of any CFC-induced ozone depletion would be an increase in the incidence of skin cancer as the flux of ultra-violet radiation reaching the Earth's surface rose. Thus in this case the anthropogenic release of CO_2 may counteract the deleterious effects of the anthropogenic release of another pollutant

4

Past climatic changes

4.1 Recent climatic changes in context

Having described the nature of recent climatic changes, in particular those which have occurred during the twentieth century, it is important to place these in the context of the longer history of the Earth's changing climate. As indicated in Section 3.1, the climatic changes that have occurred in recent times have amounted to a rise in mean annual temperature of about 0.5°C from the end of the nineteenth century up to about the 1940s, with a subsequent decline of about 0.3°C. As this chapter will demonstrate, any species which has been continuously present in a given area since early Post-glacial times will have had to withstand climatic changes of much greater amplitude (in both directions) than these. On this basis there would appear to be little likelihood of climatic changes of the magnitude of recent times bringing about the extinction of a species at a particular site. However, the fact is that comparatively small climatic changes can now produce much more significant biological effects than the larger changes which have occurred in the past as in the modern world many species of animals and plants face other environmental threats to their survival, largely as a result of man's influence on their habitat. In consequence the additional stress of a climatic deterioration may now tip the balance between survival and extinction.

A small change in mean annual temperature, such as occurred in the first half of the twentieth century, may result in a significantly large change in the number of days with a temperature above some critical biological threshold. For example, as described in Section 6.2, the 0.5°C rise in mean annual temperature from the end of the nineteenth century up to about 1940 produced an extension of the growing season of plants in England (as determined by the duration of temperatures above 6°C) of about 2 weeks. Furthermore, as pointed out in Section 3.2, although the magnitude of recent climatic changes as expressed in terms of mean annual temperature appears to be small, these changes have also been accompanied by changes in the frequency of occurrence of climatic extremes which have great biological significance (see, for example, Crisp 1964 and Dobinson & Richards 1964 for an account of the biological effects of the extreme British winter of 1962/3 and Hearn & Gilbert's 1977 report on the impact of the 1976 drought in England and Wales).

4.2 Our knowledge of past climatic changes

Biological responses to climatic change have provided an important means of elucidating the long history of past climates. In particular, the interpretation of changes in the abundance and distribution of fossil organisms has been an important technique in palaeoclimatology. If fossil remains can be identified with modern species whose tolerance limits are known, then deductions concerning the prevailing climatic conditions (especially temperature) in earlier times may be possible. A potential difficulty is that, although a fossil species may be morphologically identical with a recent species, one can not be sure that natural selection and genetic divergence have not produced changes in its tolerance limits in the intervening period. For this reason it is unwise to make climatic deductions from the distribution or abundance of a single species, although mutually corroborated evidence from several species is more acceptable.

The actual fossils used may be whole organisms, as in the case of trees preserved in peat bogs (Lamb 1964a); insects, most notably beetles (Coleoptera) (Coope 1970, 1979, Coope, Morgan & Osborne 1971); molluscs (Sparks & West 1972, Kerney, Brown & Chandler 1964); or marine organisms with hard tests (see Ericson & Wollin 1966) such as radiolarians, foraminiferans (Steuerwald & Clark 1972) and coccoliths (McIntyre 1967). But probably the most information on past climates has been obtained from palynology or the study of the abundance and ratio of pollen grains which have a highly species-specific external structure and which may be found in lake bed sediments, peat or soil in a sequence which reflects past changes in vegetation. The techniques of pollen analysis have been used extensively in the British Isles, most notably by Sir Harry Godwin and his co-workers at Cambridge (Godwin 1975, Walker & West 1970, West 1968; see also Faegri & Iversen 1975 for a general account), but have also yielded valuable information on the past climate of North America (Davis 1961, 1963), Africa (Van Zinderen Bakker 1969) and other parts of the world. Pollen data are usually presented as the percentage occurrence of individual species at various levels in a deposit, often in the form of a curve showing the rise and fall in the species' abundance as the vegetation changes over time. Particular pollen zones, delimited by sudden changes in the relative abundance of one or other species, can be recognized as consistent features from sample to sample. Knowledge of the tolerance limits of species represented in the samples allows deductions about the prevailing climatic conditions to be made. In North America, a more sophisticated approach has been adopted by Webb and Bryson (1972) who used multivariate statistical analysis to correlate pollen data with recent climatic data. The relationships thus revealed could then be extrapolated back in time to produce direct estimates of climatic variables (precipitation, maximum summer temperature) rather than the general statements about prevailing climate that are usually deduced.

The methods discussed so far relate to data concerning the presence or absence of organisms or populations of organisms and so are particularly useful in the interpretation of rather large-scale climatic changes. The resolution possible depends upon such factors as the mobility and longevity of the organisms concerned; thus Coope's work on fossil beetles has shown the existence of short periods of warmth (e.g. around 43 and 13 ka BP) which are not apparent from the pollen record because their duration was too short for long-lived and relatively immobile species such as trees to take advantage of. Much higher resolution is possible by using the techniques of tree ring analysis, as in this case one can obtain annual (and potentially seasonal) data which reflect the specific response of an individual rather than a population (see, for example, La Marche 1974, Fritts 1974, 1976, Creber 1977). This method is based on the assumption that the amount of primary production available each year for the laying down of an annual growth ring will be dependent upon the prevailing environmental conditions. If climatic conditions (e.g. summer temperature or rainfall) are the main factors limiting primary production, then by correlating the width of recent rings with contemporary meteorological data, the year-to-year variability of earlier ring widths may, by extrapolation, be used to make quite precise deductions about past climates. Thus the originator of tree ring analysis, A. E. Douglass, correlated the ring width of trees (ponderosa pine, *Pinus ponderosa,* and Douglas fir, *Pseudotsuga menziesii*) growing in the dry climate of Arizona with rainfall. Factors other than climate may cause variation in ring width; for example the production of fruit in mast years may require a disproportionate amount of the annual primary production, thus resulting in a narrow ring, as may an attack by defoliating caterpillars, but climatically determined variations in ring width would tend to have more spatial continuity and to show up consistently in trees from different sites. Of course trees growing in climatically optimal conditions will not provide useful data and are said to be 'complacent'.

Long runs of climatically useful data can be obtained either from long-lived trees such as the bristlecone pine (*Pinus longaeva;* formerly *P. aristata*) of California, which can live for more than 4000 years, or by joining together the records from a number of living and dead trees (including wood from furniture, buildings and the panels of paintings), provided that the individual data sets have recognisable 'signatures' or identifiable groups of rings which can be overlapped to produce a continuous series. Such a series or 'master chronology' can then be used for dating archaeological or art historical material, in which case a consideration of the causes of the variation in the width of the rings is not relevant. This type of approach is termed 'dendrochronology' and is distinguished from 'dendroclimatology' – the use of tree rings for climatic interpretation. The accuracy of dating attainable by means of dendrochronology has enabled tree rings to be used for correcting the ^{14}C chronology used for dating organic materials (Ferguson 1970, La Marche & Harlan 1973). The techniques of dendroclimatology have reached a high level of sophistication in North America where Fritts and his colleagues at the Laboratory of Tree Ring Research, University of Arizona, have used tree ring data to construct maps of large scale pressure anomalies going back to AD 1700. To some extent the greater development of these techniques in the United States is a consequence of the fact that many of the trees there are longer-lived than those of Europe and also they exhibit greater variability in ring width due to greater extremes of climate. However, in Europe, cross-matching of wood used in building construction has produced a 1000 year oak chronology (Hollstein 1965, Huber & Giertz-Siebenlist 1969) which has not yet been fully analysed for climatic information.

As already described above, minute fossil marine organisms (radiolarians, coccoliths, foraminiferans) preserved in deep sea sediments may be used to deduce information about past ocean temperatures, provided that their thermal tolerances are known. However, a more direct indication of prevailing water temperatures may be obtained by examining the isotopic composition of the oxygen atoms that such organisms extract from the surrounding sea water and incorporate into their hard, protective tests or shells. As the ratio of the heavier ^{18}O to ^{16}O isotope changes during any change of state and as this fractionation is temperature-dependent (Urey 1947), the $^{18}O{:}^{16}O$ ratio in the carbonate of such a test is a record of the ambient temperature at the time of incorporation of the oxygen atoms (Emiliani 1955, 1966). A similar fractionation of oxygen isotopes occurs when water passes from one phase to another (e.g liquid to solid, or liquid to vapour): water molecules (H_2O) containing ^{18}O evaporate more slowly and condense more readily than those containing ^{16}O. In consequence, the $^{18}O{:}^{16}O$ ratio in the successive layers of ice in cores taken from polar regions indicates the temperature prevailing at progressively earlier times, the deeper the layer (Dansgaard 1964) Oxygen isotope measurements from ice cores taken from Camp Century (77°N 56°W) in north-western Greenland (Dansgaard, Johnsen, Clausen & Longway, 1971) and from Byrd Station (80°S 120°W) in Antarctica (Epstein, Sharp & Gow 1970) indicate that the onset and end of the last glaciation were contemporary in both hemispheres.

The Earth sciences also contribute to our knowledge of past climatic changes, although often the data that they provide are not of high resolution: landscape features such as moraines of former glaciers and the polygonal patterning of earth cracks associated with frost action are indicative of the former extent of ice sheets. The position of early snowlines may be deduced by investigating the location of corries or cirques which are characteristic of the lower limit of permanent snow cover where alternate freezing and melting produces a bowl-shaped depression which is progressively eroded back. Hastenrath (1971) conducted a survey of the cirques along the Rockies and the Andes and was able to plot the extent of the ice age snowline from 37°N to 35°S. Other features such as raised river terraces or strandlines of lakes (which may be dateable by means of radiocarbon techniques) are indicative of earlier periods of high rainfall. Thus it appears that North American lakes were highest at the time of the northern hemisphere glacial maxima (Broecker &

Orr 1958, Flint & Gale 1958) whilst it seems that the pluvial (high rainfall) periods of Africa, registered in the raised strandlines of the great African lakes such as Lake Chad and those of East Africa, can not be equated with the glaciations manifest in higher latitudes (Butzer, Isaac, Richardson & Washbourn-Kamau 1972). Turning to more recent times, annual information on the summer monsoon rainfall of the mountains of Ethiopia is contained in the records of the Nile floods (the Blue Nile drains the Ethiopian highlands) which have been maintained since 3100 BC (Bell 1970, Riehl & Meitin 1980).

Historical records of such climatically related natural phenomena are extremely useful in deducing the record of climatic change in the recent past. For example, there is a complete series of the dates of freezing of Lake Suwa (36°N 138°E) in Japan going back to 1443 (Arakawa 1954) which has been used to derive the probable mean temperatures of each winter (December, January and February) in Tokyo, 145 km to the east (Gray 1974, 1975). Similarly, average winter temperatures in the Netherlands can be deduced from the record of the dates of opening and closing of the Haarlem–Leiden Canal from 1634 to 1839 (De Vries 1977). There is a long record of the date of opening of the port of Riga (57°N 24°E) from 1530 (Betin & Preobazensky 1959, Speerschneider 1915) which reflects the persistence of ice on the Baltic Sea and is related to the spring warming of the Eurasian continent. In addition to long records of individual events such as these, reports of a more qualitative nature have been used to derive histories of the climate in various parts of the world. Many such compilations are derived from a variety of historical sources such as diaries, annals, letters, chronicles, whose reliability may be questionable. For example, it is now known that the 1000 year record of the climate of Iceland compiled by Thoroddsen (1916) and used by Koch (1945) and Bergthórsson (1969) includes a certain amount of data from unreliable sources (Bell & Ogilvie 1978). More recent compilations, such as that of Alexandre (1976, 1977) for the climate of Belgium and surrounding areas are much more rigorous in their use of verifiable primary (i.e. contemporary) sources (see also Ingram, Underhill & Wigley 1978). Although some historical records used in the interpretation of past climates bear a direct relationship to meteorological parameters (e.g. the dates of freezing of bodies of water), other data are rather more indirectly indicative of climate and are termed 'proxy data'. Examples include the annual records of grain harvests and wine vintage dates (Le Roy Ladurie 1972) and of parish tithes (a tax of one-tenth of the annual produce of the land taken to support the church). From the seventeenth century onward, however, actual meteorological data become more and more abundant (although the rain gauge and weather vane had been in use from much earlier times) with the invention of the fluid thermometer in 1611 and the barometer in 1643.

4.3   The history of climate before the present interglacial

At various times in its history the location of the land masses in high latitudes resulting from the effects of continental drift has, as a result of consequent effects on the Earth's albedo (see Sec. 3.4) predisposed the planet to a series of glaciations. For example from various parts of the world, including Greenland, Australia, and parts of Africa and Asia, there is evidence of extensive glaciation some 800–600 Ma ago. The penultimate era of ice ages occurred 250–300 Ma ago at the end of the Carboniferous and during the early Permian. There was a period of cooling at the end of the Cretaceous about 70 Ma ago and this episode may be associated with the mass extinction of dinosaurs and certain groups of marine organisms (ammonites and belemnites) at this time, although other causes, ranging from the collision of an asteroid with the Earth, through pollution from the products of a period of intense volcanic activity to the effect of radiation from a supernova stellar explosion, have also been invoked. There was a subsequent recovery until a period at the end of the Miocene and through the Pliocene (about 10 Ma ago) when the cold intensified to result in the Pleistocene ice ages of the last 1 Ma or so.

Recent evidence, notably from deep sea sediment cores (Shackleton & Opdyke 1976; Hays, Imbrie & Shackleton 1976) indicates that there have been some ten glacial maxima, equivalent to the last ice age, in the past 1 Ma. This 100 ka periodicity coincides with the 100 ka cycle in the ellipticity of the Earth's orbit (affecting the distance from the Sun) and is evidence of the importance of orbital variations in climatic change as suggested by Milankovitch (Milankovitch 1930 – see Sec. 3.4). The naming of these glaciations has become more confused as new evidence overturned the original hypothesis of only four glaciations (named the Günz, Mindel, Riss and Würm), and now the nomenclature used varies from country to country. The penultimate glaciation (equivalent to the Riss) is termed the Wolstonian in Britain, the Illinois in North America and the Saale in northern Europe. This was followed by the Ipswichian (Britain), Sangamon (North America) or Eemian (Europe) interglacial, with the last glaciation being named respectively the Devensian, the Wisconsin and the Würm or Weichselian. There is linguistic unanimity that we are now in the Flandrian interglacial.

The penultimate interglacial began about 125 ka ago and for the first time Britain was separated from continental Europe as the sea level rose (Destombes, Shephard-Thorne & Redding 1975). The occurrence of hippopotamus bones in the East Anglian region of Britain suggests that this interglacial might have been warmer than the present one. The warmest part of the Ipswichian interglacial only lasted about 10 ka and then there was a rapid extension of ice over the land. Warm and cold periods oscillated with a general downward trend in temperature culminating in the onset of the Devensian glaciation about 70 ka ago. There were a number of warmer interstadials but the glaciation reached its height about 17–20 ka ago. In

Britain the ice sheet extended to north Norfolk but never penetrated as far south as the Wolstonian ice mass, which reached north London. In consequence there was no complete extinction of the biota during the Devensian and some species were able to maintain themselves in refuges in southern Britain or Ireland. In America plants and animals progressively moved southwards in the face of the advancing ice, but in continental Europe the east–west orientation of the mountain ranges (in contrast to the north–south arrangement of the American topography) acted as a barrier to such a retreat, as well as delaying the subsequent re-advance of species trapped behind such features as the Alps.

4.4　The postglacial

As temperatures began to rise (Fig. 4.1) and the ice melted, tundra-type vegetation dominated by grasses and sedges, dwarf birch (*Betula nana*), arctic willow (*Salix herbacea*) and the arctic-alpine flowering plant mountain avens (*Dryas octopetala*) spread northwards in Europe and into Britain which, owing to the lowering of the sea, was joined to the continent. Between 12 and 11 ka BP there was a particularly warm period called the Late Glacial Interstadial or Allerød period (after a site in Denmark where it was first recognised) when summer temperatures in England approached or perhaps exceeded those of today. Birch trees (*Betula pubescens* and *B. pendula*) and poplar (*Populus tremula*) dominated the vegetation. This warm period is also evidenced by a distinct change in the species composition of ocean sediments in the north Atlantic (Broecker, Ewing & Heezen 1960). There was a sudden reverse about 10.8 ka BP when temperatures dropped and small glaciers reformed in the English Lake District and the Scottish ice re-advanced. This post-Allerød period lasted for 600–700 years during which tundra vegetation once more became dominant in northern England. Temperatures began to rise again about 10.2 ka BP and birches spread throughout much of Britain with pine (*Pinus* spp.) dominant in the south-east. On continental Europe the ice continued to withdraw and had contracted to its present extent by about 8.5 ka BP. In North America the massive and continuous ice sheet ablated more slowly and it was not until about 7 ka BP that the modern boundary of ice cover was attained.

　　Britain was free of ice by about 10 ka BP and was joined to Europe by a land connection across the North Sea allowing free immigration of terrestrial plants and animals. By about 7000 BP the rise in sea level brought about by the melting of the ice had cut off this land connection and Britain was an island. The British fauna and flora thus consist essentially of those species which could reach Britain in the comparatively short period during which free immigration via this land connection was possible. Clearly, after Britain was cut off from the mainland, immigration by actively flying species would have continued as would have passive colonisation by animals and plants with

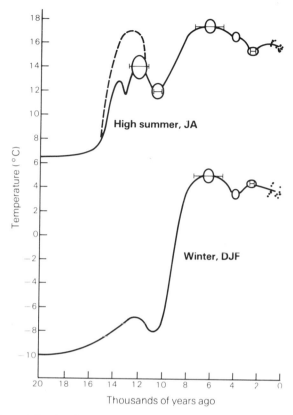

Figure 4.1 Central England temperatures over the past 20 ka. Whole lines indicate the course of the 1000 year average temperatures derived mainly from botanical evidence and the broken line represents more detailed information on summer temperatures provided by the beetle fauna (Coope, Morgan & Osborne, 1971). Oval plots indicate the range within which the mean value must lie and represent data for distinctive regimes which appear to have lasted for a period indicated by the horizontal bars. In the last millenium dots indicate temperature values derived from analysis of historical records and actual thermometer readings. JA = July and August; DJF = December, January and February. Redrawn from Lamb (1974).

wind-blown spores, seeds and other propagules, but the rate of colonisation would have been severely reduced after 7000 BP. This explains the widely remarked reduced diversity of the fauna and flora of Britain and Ireland as compared with that of Europe, a phenomenon which is particularly apparent when considering less mobile terrestrial organisms which one would have expected to have dispersed rather slowly from their ice age refuges. For example, Table 4.1 illustrates how poorly the Amphibia are represented in the British as against the European fauna, with Ireland being even more impoverished. A number of species known from fossil evidence to have been

Table 4.1 Number of amphibian species in Europe, Britain and
              Ireland.

Group	No. spp. in Europe	No. spp. in Britain	No. spp. in Ireland
Anura (toads and frogs)	24	3	2
Urodela (newts and salamanders)	19	3	1
Apoda	1	0	0

present in Britain during earlier interglacials did not make the return journey
after the last glaciation in time before the melting of the ice cut Britain off from
the Continent. Examples include the fallow deer (*Dama dama*), which was re-
introduced by the Romans, and trees such as spruce (*Picea abies*), re-
introduced *c.* AD 1500, and fir (*Abies* spp.), re-introduced *c.* AD 1600. It
would seem likely that the first species to reach Britain would be cold-tolerant
species and that it is forms with higher thermal requirements that are lacking
from the contemporary fauna and flora. Some of the original colonisers may
have become extinct as temperatures continued to rise during the Postglacial
(reaching annual means of about 2°C higher than at present in the Postglacial
Climatic Optimum of 7000–5000 BP), so that the surviving members of the
British fauna and flora would be those with wide tolerance limits, able to
withstand temperatures both lower and higher than those of today (i.e.
eurythermal as against stenothermal species).

Temperatures continued to rise in the Boreal Period of 9000–8000 BP when
pines became abundant all over England. Hazel (*Corylus avellana*) spread
northwards followed by elm (*Ulmus glabra*) and oak (*Quercus robur* and *Q.
petraea*), which became more abundant at the expense of the birch and the
pine. In North America alder (*Alnus*) colonised recently deglaciated ground
followed by spruce (*Picea sitchensis*) then western hemlock (*Tsuga
heterophylla*) and the more warmth-loving redwood (*Sequoia*), oak and
Douglas fir (*Pseudotsuga menziesii*). The Boreal Period led on to the Atlantic
Period which culminated in the Postglacial Climatic Optimum of broadly
7000–5000 BP when mean temperatures were about 2–3°C above present
ones. In Sweden, hazel spread further north than its present distribution,
indicating summer temperatures perhaps 2.4°C warmer than at the beginning
of the present century (Andersson 1902). In Britain and Europe the warmth-
loving lime (*Tilia cordata*) (one of our most thermophilous native trees), and
elm were present and are now thought to have dominated the forest together
with oak and, in wetter places, alder (*Alnus glutinosa*). Pine had practically
disappeared except in the northern Scottish Highlands, although birch still
persisted in the north-east. Ivy (*Hedera helix*), which is very sensitive to low
winter temperature, was abundant in Scandinavia, indicating that this was a
period of markedly oceanic climate, as also evidenced by the growth of peat
bogs in Britain. In addition, tree lines were generally about 200–300 m higher
and trees grew in the Hebrides and Iceland.

At the end of this period there was a sharp oscillation to a colder climate which lasted for 300–500 years and signalled the change from the warm, wet conditions of the Atlantic Period to the generally drier but more variable Sub-Boreal Period which lasted from about 5000 to 3000 BP. Elm and lime declined whilst oak became more dominant. The sudden and simultaneous decline of the elm throughout Europe (although different species were involved in different areas) at about 5000 BP is a characteristic feature of many pollen profiles. Doubts have been expressed concerning a purely climatic explanation, however, as in Britain there is no clear indication of a decline of frost-sensitive species like the ivy and mistletoe at this time. However, this may just indicate that even in this colder regime frosts were no more severe than in recent times. In fact (H. H. Lamb, personal communication) the cold oscillation around 5500–5000 BP was almost certainly less severe than the Little Ice Age of AD 1550–1850. The rise of the Neolithic culture and the spread of agriculture from about 6000–5000 BP onwards means that the influence of man on the environment becomes increasingly important, and it has been suggested that the elm decline was associated with the gathering of elm leaves as fodder for stalled cattle.

About 4000 BP world sea levels reached their highest point (about 2–7 m above present levels) as a result of the melting of the ice during the Postglacial Climatic Optimum. However, from roughly 3000 BP there was a marked deterioration worldwide into the cold, wet Sub-Atlantic Period characterised by the advance of glaciers (there had been a temporary advance around 5200 BP), regrowth (producing characteristic 'recurrence surfaces') of peat bogs and lowering of tree lines.

4.5 Historical times

From about 500 BC (i.e. 2500 BP), when the climate was colder than at present, there was a gradual recovery (which, from proxy data, appears also to have encompassed much of North America) to the greater warmth and drier conditions which prevailed during Roman times, with a peak in the fourth century AD when it may have been warmer than in the twentieth century. At this time the warmth-loving vine was cultivated in northern France and southern England. There was a downturn in temperature between AD 400 and 800–900 and conditions were at times (e.g. the 580s) much wetter. Subsequent warming produced the Little Optimum or mediaeval warm period, the timing of which varied around the hemisphere but was broadly between AD 900 and 1300. In England and north-western Europe summer temperatures were about 1°C higher than at present (see Fig. 4.2) and winters were mostly mild and wet, although there were still occasional severe ones. Sea levels rose, especially around the North Sea (about AD 1000 an inlet like a fjord brought the sea up to Norwich, now about 30 km from the East Anglian coast) and in the second half of the thirteenth century a large area of the Netherlands was

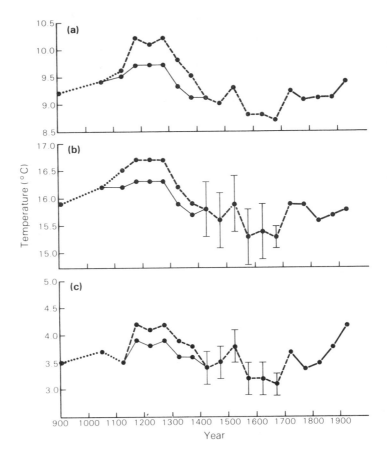

Figure 4.2 Central England temperatures: 50 year averages from AD 800. (a) Annual figures; (b) high summer (July and August); (c) winter (December, January and February). Vertical bars indicate ±three standard errors. ●——● directly observed values; ●——●unadjusted values derived from statistical relationships with the frequencies of reported seasonal weather characteristics, with no correction for reporting bias; ●----● preferred values adjusted to fit botanical indications, e.g. crop yields, wine harvests; ●·····● connects points corresponding to 100–200 year (rather than 50 year) means indicated by sparse data. Redrawn from Lamb (1965).

inundated to produce the Zuyder Zee. This was a period of enhanced oceanic influence produced by an increased frequency of westerly winds. Consequently, in the 1200s the plains of Iowa began to be affected by a drought due to the more extensive rain shadow resulting in the lee of the Rocky Mountains, and oak woodland gave way to more arid grassland. Further north, in Greenland (which was settled in 987, Iceland having been colonised in 870) temperatures peaked in the eleventh century, 100–200 years earlier than in Europe and North America. Sea temperatures in this area were also higher (sea ice was not mentioned by the early Viking explorers but had been recorded by Irish missionaries between the sixth and the eighth centuries) and cod were abundant off the western coast of Greenland.

The biological response to this mediaeval warm period is described in historical records and is evidenced by the remains of higher tree lines and traces of tillage, which in southern Scotland, northern England and Devon can be detected about 150–200 m higher than the present limits. Crop production was generally more extensive than today with oats and barley being grown in Iceland and the dry Breckland of eastern England being successfully cultivated. Many productive vineyards (*Domesday book*, William the Conqueror's survey of the lands of England compiled in 1086 as an aid to taxation, records 38 vineyards in addition to those of the king) were maintained without benefit of protecting walls in England south of about 53°N, which is approximately 500 km further north than the modern limit of commercial vineyards on the Continent. This indicates a generally milder climate with warm summers as the northern limit of vine cultivation is approximately defined by the 18°C July isotherm. However, at the edge of its range the vine is particularly susceptible to cool temperatures at either end of the growing season, so we can deduce that at this time England enjoyed warm, sunny autumns and an absence of late spring frosts. The monastic revival following the Norman conquest added a new impetus to the cultivation of the vine in England and after 1100 the area of land under vines increased markedly. Considerable quantities of wine were required for the large populations of workers and their families which grew up around monasteries, as well as for the monks themselves, who were far from abstemious. Even such an ascetic order as the Carthusians were permitted a pint of wine with their main meal of the day and the Benedictines of Battle Abbey in Sussex were each allowed a gallon of wine a day, an allocation which was increased should a monk fall ill (Seward 1979). However, a series of poor summers in the mid-thirteenth century dealt a severe blow to the English vineyards and more wine began to be imported from France (interestingly enough, English vineyards continued to flourish after the marriage of Henry II to Eleanor of Aquitaine in 1152 which added the vineyards of Bordeaux to the English crown).

This heralded a more general downturn of the climate with a series of bad summers throughout Europe in 1310–25 (with great famines in England in 1311–17). These events, repeated in 1345–48, produced widespread desertion of farms and decline and abandonment of villages even before the marked demographic changes resulting from the Black Death which (Ziegler 1969) spread across Europe between 1348 and 1350. In Iceland, where the climax of the warm epoch was earlier, grain growing was largely abandoned between the thirteenth and fourteenth centuries, not to be resumed until the 1920s. (It has again declined greatly since 1960.) Similarly, in much of North America the cooling started early in the thirteenth century.

There was similar abandonment of settlements and starvation in the 1430s when throughout Europe there was a virtually unparalleled run of severe winters, probably attributable to a persistent anticyclone over Scandinavia, with great year-to-year variability in summer weather (droughts alternating with years of high rainfall) indicating a greater frequency of blocking (see Sec. 3.2). There was a period of some recovery with better summers and warm winters in England between 1500 and 1540, although much of Europe continued cold. In the late sixteenth century, however, there was a further deterioration with wet summers and cold winters. The period between 1550 and 1850 is often referred to as the Little Ice Age and is registered in the tree ring records from the White Mountains (119°W) in North America (La Marche 1974) as well as the European historical records as a period of marked climatic cooling.

The worldwide increase in the area of snow and ice cover during the Little Ice Age has been estimated as 10–15% of the anomaly that developed in the Devensian ice age, although its duration was only 2–3% of the last glaciation (Lamb 1964b). During this period the mean temperatures were low in all seasons by comparison with the present century but there was also great year-to-year variability attributable to blocking. For example, in the long record (from 1530) of the extent of ice on the Baltic Sea given by the dates of opening of the port of Riga (see Sec. 4.2) the two years (before 1975) with least ice and the year with most ice all fell within the same decade, the 1650s. There was a great increase in storminess during this period, especially around the North Sea, e.g. in 1530, 1570, 1634, 1697 and the 'Great Storm' which passed across southern England in December 1703. One of the less extreme storms climatically but of great significance historically was that of September 1588 which destroyed many of the ships of the Spanish Armada and was recorded in the inscription *'Flavit et dissipati sunt'* ('It blew and they were scattered') on the commemorative medal issued by Queen Elizabeth I. During this same period there were also notable movements of windblown sand, e.g. at the Culbin Sands (57.6°N 3.7°W), south of the Moray Firth, Scotland, in 1694. The increased windiness also contributed to the destruction of the vegetation cover of the Breckland in East Anglia.

It is only from the mid-seventeenth century that we have actual instrumental observations of meteorological variables and the longest continuous record of temperatures in existence is the central England series from 1659 to the present day (Manley 1974 – see Fig. 4.3). This is a standardised series of monthly mean temperatures for a typical lowland site in central England compiled from overlapping runs of years of observations from different places between Lancashire, East Anglia and the Thames Valley. The series is continued today as the mean of the values for a triangle of stations: Nelson, Lancashire; Botanic Gardens, Cambridge; and Ross-on-Wye. Landsberg, Yu and Hwang (1968) have constructed a similar temperature series for the eastern USA going back to 1738. Similar averages have been used to construct a continuous rainfall series for England and Wales from 1727 (Nicholas & Glasspoole 1931, Wales-Smith 1971). Although Manley's central England temperature series does not cover the first 100 years of the Little Ice Age, it includes one notable winter, that of 1683/4, known as the 'long frost', which appears to be the

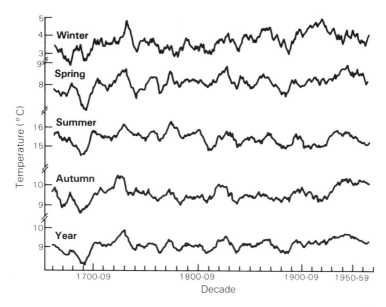

Figure 4.3 Manley central England temperature series 1659–1973. Average surface air temperatures at typical lowland sites in central England: 10 year running means. Redrawn from Manley (1974).

coldest in the instrumental record for England but may have been equalled in 1607/8 also. The 1683/4 winter was associated with high pressure blocking near Iceland, rather like the extreme winter of 1962/3 (Manley 1975). With the seas probably in general colder than nowadays, in 1683/4 ice belts some miles in width formed on the coasts of the English Channel, East Anglia and the Netherlands, and a celebrated Frost Fair was held on the Thames at London. The 1690s constituted a particularly extreme run of years, the effects of which were especially marked in those areas that were marginal for the growth of crops. In the upland parishes of Scotland, for example, the valley bottoms were so wet in the seventeenth century that ploughing spread up the steep hillsides, increasing the vulnerability of the communities to the effects of the cold, wet summers of 1693–1700. In consequence about one-third of the population died of starvation and the resultant demoralisation of the Scottish people had a profound influence on the debate leading up to the union of the English and Scottish parliaments which eventuated in 1707.

The cause(s) of the Little Ice Age is, as with all climatic changes, difficult to determine, but there is an interesting correlation between the worst of the period and an apparent, and unprecedented, gap in the naked-eye sunspot record between 1645 and 1715. The suggestion that a decrease in the solar constant during this 'Maunder Minimum' (Eddy 1976) may have resulted in the Little Ice Age is attractive, but computer modelling (Robock 1979) indicates that the effect of volcanic dust produces a better simulation of climatic change during, and after, this period. Whatever the causal mechanism, the ubiquity of this downturn in temperature is indicative of a sudden change in some factor external to the ocean–atmosphere system.

The recovery from the cold of the Little Ice Age seems to have begun in the early eighteenth century: all seasons during the 1730s had temperatures equivalent to or even higher than those of the warmest decades of the present century. This decade, however, was only a short-lived glimpse of what the twentieth century held in store and was brought to a close by the severe winter of 1739/40. Temperatures dropped back again, especially winter temperatures; however, from about 1720 to 1810 summer temperatures tended to be warmer than the average for the present century. The winters of 1812–20 during the childhood of Charles Dickens (1812–70) were notably cold and clearly influenced that author's conception of Christmas weather. 1816 in particular was one of the worst years in Europe and the United States, with a wet, cold summer following a hard winter. This 'year without a summer' was probably the result of the eruption of the volcano Tambora in Indonesia in 1815. Reliable barometric pressure records for western and central Europe are available in sufficient quantities to enable the reconstruction of monthly mean pressure charts of the January and July atmospheric circulation back to 1750 (Lamb & Johnson 1966) and these indicate that winters in the period 1790 to 1829 were marked by a weakening of the westerly winds and a tendency to blocking, a pattern that again became prevalent in the 1940s and since about 1960.

The effects of the gradual rise in temperature since the eighteenth century became manifest in the 1850s with the beginning of the recession of the Alpine glaciers, although from 1875 to about 1890 there was a return to cooler conditions with Dickensian winters due to high pressure blocking over Scandinavia directing polar air, and water, southwards (in 1888 the Arctic ice advanced to near the Faeroe Islands, 62°N 7°W).

5
Mesoclimate – the interaction of climate and habitat features

5.1 Introduction

Subsequent chapters of this book will consider the variety of ways in which the natural fauna and flora have responded to the twentieth century climatic changes which, as described in Chapter 3, have been identified in terms of trends in weather variables such as temperature, or the frequency of weather extremes such as hard winters or drought summers. Chapter 3 also explained how these climatic changes are associated with large scale features of the climate, especially those connected with the strength and location of features of the circumpolar vortex. These aspects of the synoptic climate may bear little relationship to the actual conditions experienced by an animal or plant in its natural habitat. The conditions pertaining in an organism's immediate environment, whether, for example, in a grass tussock, beneath a layer of leaf litter, or in a tree-hole nest, can be accurately determined and are said to constitute the organism's microclimate (Geiger 1966, Cloudsley-Thompson 1967). However, there are larger scale habitat features, e.g. the slope of a hillside, the percentage vegetation cover or stand height, which interact with climatic variables. This interaction may produce a modification of the climate actually experienced by an organism that is on a rather broader scale than that usually covered by the term microclimate, and accordingly the term 'mesoclimate' is more appropriate in this context.

5.2 Aspect

One important environmental feature that has a significant effect on the mesoclimate of a particular site is its aspect, or the direction in which it slopes, particularly with reference to the position of the Sun. In the northern hemisphere, south-facing slopes receive more sunshine than north-facing ones and so phenological events (see Sec. 6.5) occur earlier there. Pollard's (1975) work on the Roman or edible snail (*Helix pomatia*) indicates the importance of aspect in determining the distribution of species such as this which have high thermal requirements. As its name suggests, this snail was introduced into Britain by the Romans and now has a limited distribution in the south of the country. As it has a high calcium requirement owing to the large quantities of calcium carbonate incorporated into its shell it is restricted to calcareous soils. The main centres of its distribution are the chalk of the North Downs, the chalk of Hertfordshire and the limestone of Gloucestershire. Pollard examined the 195 site records of *H. pomatia* for which six-figure grid references were available and found that the species generally occurred on steeply sloping countryside with a preferred southerly (Surrey) or south-westerly (Kent) aspect. The preference differences between these two counties may merely reflect the differing aspect of the scarp slopes on which many of the snail populations occur. In Gloucestershire the scarp faces north-west and there is no statistically significant preferred aspect, most sites occurring in valleys on the dip slope. In Hertfordshire the snail sites are all to the south of the north-west-facing scarp but there is still a preference for southerly or south-westerly aspect out of the wide choice available. A similar preference was found to apply at the local level when the occurrence of snails was examined with reference to the aspect of roadside banks in a Hertfordshire lane. Pollard's data show that 82% of snails occurred on banks with a southerly or south-easterly aspect. Pollard draws attention to the scarcity of records from the calcareous soils of the South Downs and Chilterns and suggests that, for the former, the lack of a major south-facing equivalent to the scarp of the North Downs may be a contributory factor. However, it is difficult to reconcile this with the presence of snail populations on the chalk of Hertfordshire and the limestone of Gloucestershire.

In general, butterflies only fly in sunshine (see Sec. 6.7) and so are often more abundant at sites with a southern aspect where insolation is higher. Muggleton (1973) reports that the chalkhill blue (*Lysandra coridon*), Adonis blue (*Lysandra bellargus*) and small blue (*Cupido minimus*) are usually found on slopes with aspects between southerly and westerly. The recently extinct large blue (*Maculinea arion*) used only to be found on south-east- and south-west-facing slopes and Muggleton draws attention to the fact that this species was reputed to fly only between 10.00 and 12.00 hours and again at around 16.00 hours, at which times the sun would be shining on either the south-east- or the south-west-facing sites. The black hairstreak butterfly (*Strymonidia pruni*) lays its eggs on blackthorn (*Prunus spinosa*) trees in woods of the East Midlands forest belt. It generally prefers sheltered, but unshaded, sites such as glades, rides and sheltered edges. Although there was no preference for a particular aspect when rides were considered, Thomas (1976) found that the linear breeding areas along hedgerows and woodland edges generally had a south- or south-west-facing aspect.

One would expect aspect to be of particular significance to poikilotherms as their metabolism is directly affected by temperature, but N. W. Moore (1962) reported that the Dartford warbler (*Sylvia undata*), a bird with a very restricted distribution in southern England, favours sheltered areas and south-facing slopes for breeding. This may indicate an effect of aspect on the availability of the bird's insect food (see Sec. 9.2), although Bibby and Tubbs (1975) were unable to detect any clear preference for any particular aspect in this species.

A corollary of the increased sunshine on south-facing slopes is that certain markedly oceanic organisms prefer a north-facing slope where the sun has least influence. For example, Ratcliffe (1968) has shown that the so-called Atlantic species of bryophytes, which require an atmosphere saturated with water vapour, are most abundant and attain the greatest luxuriance on slopes with an aspect facing between north-west and due east. The steeper the slope, the greater is the shading effect and the richest localities have an angle of incline greater than 25°. The filmy ferns (or Hymenophyllaceae), such as *Hymenophyllum tunbrigense* and *H. wilsonii,* have similar requirements to Atlantic bryophytes and similarly prefer sites with a northerly aspect.

5.3 Thermally favourable habitats

The example of the Roman snail (*Helix pomatia*) described in Section 5.2 illustrates the general proposition that the favourable combination of a dry, freely draining substrate and south-facing slopes means that the mesoclimate of chalk hills provides an optimal habitat for thermophilous species. As well as facilitating the activity of animal species, this also applies to such plants as the horseshoe vetch (*Hippocrepis comosa*) (Fearn 1973), the food plant of the Adonis blue butterfly (*Lysandra bellargus*), and the pasque flower (*Pulsatilla vulgaris*), a species whose distribution is limited by the 16°C mean July isotherm and so in England is confined to south-facing slopes with high insolation on dry calcareous sites in the east (Wells & Barling 1971).

Other types of substrate similarly constitute thermally favourable environments, either because they heat up rapidly or because they lack shading vegetation, and so provide suitable habitats for species with high thermal requirements. There are a number of examples of this among the spiders (I am grateful to Dr Eric Duffey for directing my attention to this phenomenon), especially the lycosids or wolf spiders which require high temperatures for the development of their egg cocoons which the females carry around (see, for example, Norgaard 1951).

The lycosid *Arctosa perita*, although widely distributed in the British Isles, is generally found on substrates which are relatively free of vegetation and which in consequence warm up rapidly. These include slag heaps, sandhills and dry heaths where the spider is frequently found basking in the sun. *Pardosa hortensis* is a more southern species and is often found in sand and gravel pits, as is *Xerolycosa nemoralis*, which has a very limited distribution in the south. This latter species occurs on sparsely vegetated heathlands in Surrey and is also found basking in woodland clearings. Two species of salticid or jumping spider, *Sitticus rupicola* and *Euophrys browningi*, also have very limited distributions in the south, where they occur mainly on shingle beds which heat up rapidly. The former was originally only recorded from shingle at two localities, namely Shoreham in Sussex and Hayling Island in Hampshire (Locket & Millidge 1951) but was subsequently found to occur on shingle at a number of coastal sites in the south and south-east, from Bridgewater Bay in Somerset, along the coast of Hampshire, around the Isle of Wight and up to Norfolk (Locket, Millidge & Merrett 1974). *Euophrys browningi* is only found on shingle beaches, where it often inhabits empty whelk shells, in the south-east. Its most celebrated locality is Shingle Street in Suffolk, but it is also found in Essex at Colne Point, Bradwell and Walton-on-the-Naze, near Faversham in Kent and as far north as Blakeney Point in Norfolk.

Thermally favourable environments are important for poikilothermic vertebrates, especially for their successful reproduction and development. The natterjack toad (*Bufo calamita*), for example, requires high temperatures for the successful development of its larvae or tadpoles and so breeds late in the season in comparison with other amphibians (mid-April–June) in shallow, exposed pools which heat up rapidly and maintain high temperatures. Laboratory experiments indicate that temperatures above 25°C are optimal for larval development in this species (Mathias 1971, quoted in Beebee 1979). For this reason the species is restricted to ponds in sandy habitats where the soil has a low nutrient status for plant growth resulting in an absence of high, shading vegetation. In consequence high water temperatures are attained, but, as a corollary, unusually hot summers, such as those of 1975 and 1976, are deleterious as the shallow breeding pools dry up before the tadpoles metamorphose: there is a fine dividing line between conditions producing a successful breeding season and those resulting in complete failure. The newts *Triturus vulgaris* (smooth newt), *T. helveticus* (palmate newt) and *T. cristatus* (crested newt) are similarly affected by summer drought which causes their breeding ponds to dry up before the animals have successfully metamorphosed.

Bufo calamita was formerly found on coastal dunes and sandy inland heaths throughout England (Taylor 1948) but is now virtually restricted to a few coastal colonies in East Anglia and Lancashire and one remaining area of heathland in southern England (Beebee 1976). Prestt, Cooke and Corbett, (1974) chronicle the decline of the natterjack toad in different areas and from this and other surveys it is clear that the decline has been primarily due to the destruction and alteration of its habitats by the activities of man: recreational pressure and building on coastal areas, together with the draining of dune slacks, and destruction of heathland by urban and agricultural development and afforestation. Beebee (1974) showed that in general there was no direct correlation between the decline of *B. calamita* and climatic trends, although in some areas lowered rainfall appears to have been a contributory factor which became more significant in the 1970s. Rather, this is an example of a species which is restricted to a particular type of habitat for climatic reasons and in consequence has declined as that habitat has decreased in extent due to changes in land use resulting from human activities.

Another poikilothermic species, the sand lizard (*Lacerta agilis*), requires sandy substrates with no shading vegetation on which to lay its eggs and is thus similarly restricted to sand dunes and sandy lowland heath, and has also declined in recent years as these habitats have been destroyed so that it is now restricted to a few isolated heathland localities in southern England as well as one coastal sand dune population in Lancashire. However, all the remaining sites are subject to human disturbance and fire, accidental or deliberate, is another deleterious influence at the heathland sites, an influence which is predominant in years with warm summers which would normally be expected to be optimal for successful reproduction. Another relevant factor is that the patches of exposed sand (which heat up rapidly) required for egg laying were formerly provided by the close grazing activity of rabbits (*Oryctolagus cuniculus*), so the reduction in the rabbit population as a result of the introduction of myxomatosis has also played a part in the decline of the sand lizard.

As the above account indicates, it is mainly the influence of habitat factors of one kind or another that have accounted for the decline of this species, but the impact of such factors will clearly be greatest during a period of deteriorating climate. It is therefore interesting that Jackson (1978) has demonstrated that a reduction in May sunshine levels – the 6.5 h May isohel appears to correspond with the species' northern boundary in Britain – is implicated in the decrease in the isolated sand lizard population of north-western England in recent years.

Although another egg-laying reptile, the grass snake (*Natrix natrix*), frequently solves the problem of achieving the high temperatures necessary for development by laying its eggs in natural or artificial (i.e. compost heaps) piles of rotting vegetation, other species are less dependent on habitat features for the provision of the required mesoclimate owing to their viviparous habit. The female adder (*Vipera berus*), for example, moves in and out of the sun in order to keep the eggs that are developing within its body at an approximately constant temperature favourable for development. The smooth snake (*Coronella austriaca*) is an unusual example because, although viviparous like the adder and so unlikely to be limited in its distribution owing to its thermal requirements, it is restricted to dry lowland heath in southern England. However, this association may not be a direct influence of climate, but perhaps a reflection of the smooth snake's marked preference (Prestt *et al.* 1974) for reptile prey which are only present in sufficient abundance to support viable populations of such a predator in the thermally favourable environment provided by southern heathland.

5.4 Climatic refuges

The account of the habitat preferences of terrestrial poikilothermic vertebrates in the previous section indicates the significance of the mesoclimate provided by dry heathland habitat. Unfortunately this habitat is vulnerable to a number of different threats (agricultural reclamation, afforestation, mineral extraction, urban development) and its areal extent has decreased considerably at an accelerating rate. N. W. Moore (1962) has shown how the area of heathland in Dorest declined from about 30000 ha at the time of the first edition of the Ordnance Survey maps in 1811 to 23000 ha in 1896 when a second edition was produced, and only 10000 ha by 1960. The current area is estimated to be about 6000 ha (Webb & Haskins 1980). Furthermore, the remaining area of heathland has become greatly fragmented with the consequence that some of the surviving areas no longer constitute viable habitat units for certain species. This is the case with the Dartford warbler (*Sylvia undata*), which was brought to the verge of extinction by a combination of the destruction of its heathland habitat and the effects of the severe winters of 1961/2 and 1962/3 (see Sec. 9.2). This essentially non-migratory bird was formerly quite widely distributed in southern England but has become more and more restricted as the available habitat has become reduced owing to the pressure of human activities. Furthermore, the remaining heathland is now much more fragmented with a consequent reduction in the possibility of recolonisation if a particular population is wiped out after bad weather. For example, after the severe winter of 1946/7 populations of Dartford warblers were re-established in the Isle of Wight, northern Hampshire, Surrey and Sussex, all by recolonisation from the New Forest in Hampshire. However, after the 1961/2 and 1962/3 winters, by which time habitat destruction and fragmentation had proceeded further, the Surrey and north-eastern Hampshire population of about 40 pairs (Bibby & Tubbs 1975) failed to re-establish itself (see Fig. 5.1).

Another example of the interaction between adverse climatic factors and a reduction in the availability of optimal habitat is the case of the extinction, or near-extinction, of the unique *masseyi* subspecies of the silver studded blue

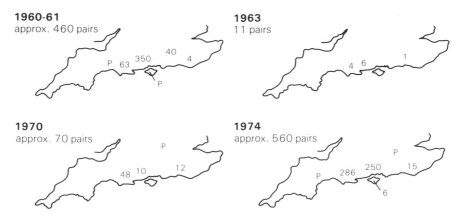

Figure 5.1 Approximate numbers and distribution of breeding Dartford warblers (*Sylvia undata*). Figures represent estimated numbers of pairs breeding in Devon, Dorset (new boundary), New Forest, Isle of Wight, north-eastern Hampshire and Surrey combined, and Sussex. P indicates numbers uncertain but less than five pairs believed present. Redrawn from Bibby and Tubbs (1975).

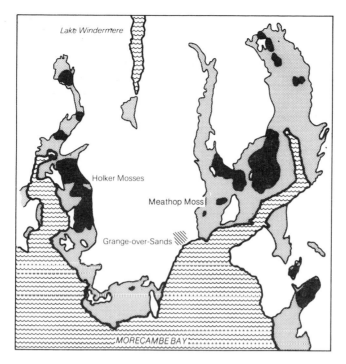

Figure 5.2 Habitat of the *masseyi* form of the silver studded blue butterfly north of Morecambe Bay: present distribution of the mosses in black, alluvial areas representing their presumed former distribution are stippled. Redrawn from Satchell (1975).

butterfly (*Plebejus argus*). The distinguishing feature of this race is that the wings of the female are coloured blue rather than the usual brown. This subspecies was restricted to the lowland mires or 'mosses' to the north of Morecambe Bay, Lancashire, especially Holker Mosses and Meathop Moss (see Fig. 5.2). It was last found in quantity in these mosses in 1921 (generally a good summer) when, in the third week of May, when the butterfly would have been present in the form of almost fully-grown caterpillars, there was a very severe late frost which wiped out the animal. This frost was clearly not unprecedented and so the question arises as to how the butterfly was able to survive previous late spring frosts. The answer appears to lie in the fact that formerly the mosses were much more extensive and included those areas of alluvium in coastal and valley areas (see Fig. 5.2) which are now tree-covered (Satchell 1965). The original reports of the 1921 frost describe how all the vegetation on the mosses was blackened up to a height of about 9 feet (3 m) while on the crags above the vegetation was hardly affected (Wright, 1942). The mosses that formerly covered these alluvial areas in the valley heads were much higher than the remaining mosses and so would have provided refuges from which, after similar frosts, the coastal mosses could be colonised again.

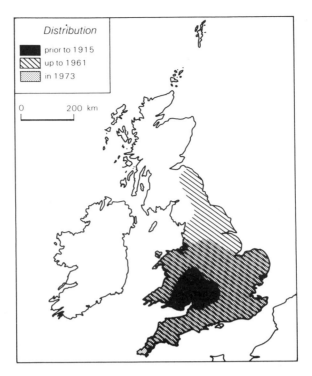

Figure 5.3   Changes in the distribution of the comma butterfly (*Polygonia c-album*).

The destruction of these habitats ensured that eventually an unseasonable frost would eliminate the butterfly. In fact, this subspecies has been recorded on two occasions since 1921, although apparently not within the last 40 years, so presumably a relict population persisted somewhere in the region. Howarth (1973) suggests that final extinction was brought about by a fire in 1941.

   The history of many species, especially animals which are more mobile, has probably involved periodic expansions of range as the climate ameliorates, with subsequent retreats to climatically favourable refuges as less optimal conditions return. This is probably the case with two other British butterfly species, the comma (*Polygonia c-album*) and the white admiral (*Ladoga camilla*). The comma is essentially a woodland species which, up to the middle of the last century, was widespread in many counties in England from Somerset up to Durham and Cumberland. It subsequently contracted its range until by 1915–20 it was confined to a core area in the West Midlands covering parts of Worcester, Gloucester, Hereford and Monmouth. Over the next few decades, however, it expanded its range throughout Wales and southern England (see Fig. 5.3), reaching Lancashire and Yorkshire again by 1950, and eventually spread as far as the Scottish border. Its range has now contracted again and it is no longer found north of the Humber. The changes in this species' distribution during the present century parallel known climatic changes, and the West Midlands area presumably constitutes a climatically favourable refuge, although the precise features of this region that contribute to the provision of an optimal mesoclimate for the comma butterfly are as yet undetermined.

The white admiral is a more continental species which, although previously widespread in the southern counties of Britain, became restricted to the New Forest in Hampshire and a few isolated woodlands in the extreme south of England and the Isle of Wight from about the middle of the last century. In the 1920s, however, it began to extend its range and was eventually found throughout the south-east from Cornwall to the Humber (see Fig. 5.4). This spread must have been associated with the twentieth century climatic amelioration as artificial attempts to extend the range of the white admiral had previously failed. Thus E. B. Ford (1957, p. 140) reports:

In 1907 or 1908 Canon Godwin liberated large numbers of the White Admiral in Wateringbury Woods but none survived. Long afterwards that locality became included within the range of the butterfly which had extended naturally, and in 1934 Canon Godwin saw at least 200

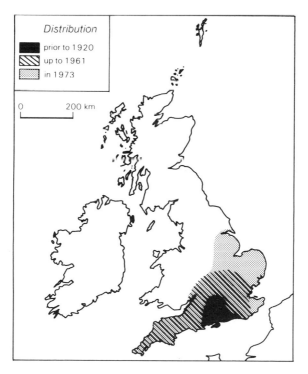

Figure 5.4 Changes in the distribution of the white admiral butterfly (*Ladoga camilla*).

specimens in the area where he had unsuccessfully attempted to establish the species in the past.

Pollard (1979b) has shown that over the period of expansion of range of the white admiral, especially from 1930 to 1942, high June temperatures increased survival by reducing the duration of the late larval and pupal stages, which are susceptible to bird predation (see Sec. 6.3). Habitat changes have also had a direct influence on the species as the abandonment of the traditional coppice management of woodland, which included the deliberate removal of the butterfly's food plant honeysuckle (*Lonicera periclymenum*) which wraps itself around trees, probably increased the availability of habitat suitable for colonisation by the species. However, the climatic influence has been predominant again since about 1960 as the species has once more been withdrawing to the centre of its range, although the good summers of 1975 and 1976 allowed numbers to build up again.

5.5   Habitat restriction at the edge of a species' range

An important point to bear in mind when considering the interactions between climate and features of the habitat in such a location as Britain, where many species are at the northern limit of their distribution, is that species on the edge of their range tend to have more specific habitat requirements. For example, the sand lizard discussed above occurs in a variety of habitats on the Continent where the warmer summers impose fewer limits on egg laying. Among the Lepidoptera, although in Britain the large blue butterfly (*Maculinea arion*) (see Section 6.7) used to feed and oviposit only on wild thyme (*Thymus drucei*), in France marjoram (*Origanum vulgare*) is also acceptable. Similarly, the British race of the swallow-tail butterfly (*Papilio machaon britannicus*) is restricted to the fens in England where the milk parsley (*Peucedanum palustre*) is almost its sole food plant. On the Continent, however, the butterfly is common in woods and meadows (often quite dry) and up to a height of 1500 m and its diet includes a wide variety of umbelliferous plants.

The same phenomenon is found in plants, especially oceanic species such as ling or heather (*Calluna vulgaris*), so characteristic of maritime heathland, which becomes increasingly restricted to forests or other habitats maintaining a relatively high atmospheric humidity in the more continental parts of its European range (Gimingham 1960). Similarly, the stemless thistle (*Cirsium acaule*) grows on hills of all aspects in south-eastern England and northern France but on the edge of its range in the Yorkshire Wolds and Derbyshire (see Section 6.2) it is confined to the warmer south-facing slopes.

6

Direct metabolic effects of climate

6.1 Introduction

The metabolic processes of all organisms are temperature-dependent (Sec. 2.1). Homeotherms (birds and mammals) are able to maintain a high and approximately constant internal body temperature and so are relatively independent of prevailing environmental temperatures. However, the physiological processes of all plants and of poikilothermic animals (together comprising the vast majority of living things) proceed at a rate dependent upon ambient temperatures. In consequence, there are diurnal (e.g. Phillipson 1962) and day-to-day variations in metabolic rates, but the annual cycle of temperature is perhaps of greatest significance in considering the biological impact of climate and climatic change. The marked seasonality of the climate in temperate regions such as the British Isles means that for perhaps half the year temperatures are below the minimum necessary for metabolic processes to proceed in many plants and animals and so such organisms either die down or overwinter in a resting form such as a pupa or egg. Climatic changes producing a small change in annual temperature may result in a significant reduction in the length of the metabolically active season. Such a reduction may affect the survival of an organism in a particular area if there is no longer a sufficient length of time between spring and autumn for the completion of the processes of development, growth and successful reproduction.

6.2 Growth and reproduction of plants

The growing season of plants is conventionally defined in terms of the number of days when the mean air temperature exceeds either 42°F (= 5.6°C) or 6°C. This threshold has particular relevance for agricultural crops but also represents the temperature above which pasture grasses actively grow and so is probably generally applicable to plants in temperate regions. As temperature decreases with altitude at a rate of about 1°F per 300 ft or 0.65°C per 100 m, a map of the average duration of the growing season, such as that for England and Wales for the period 1921–50 shown in Figure 6.1, reflects the combined effects of latitude, height above sea level and, of course, oceanicity. As the curve of the annual march of temperature for a maritime climate such as that experienced by much of Britain is rather flattened, a small change in mean annual temperature, as shown in Figure 6.2, makes a great difference to

Figure 6.1 Duration of growing season in England and Wales (in days). Redrawn from Hogg (1965).

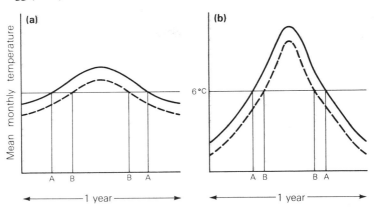

Figure 6.2 Effect of a small change in mean annual temperature on the length of the growing season. An overall decrease in mean monthly temperatures (represented by the dashed line) results in a greater reduction in the length of the growing season (A–A reduced to B–B) at a maritime site (a) than at a continental site (b).

the length of the growing season, that is that part of the curve above the 42°F or 6°C line. Thus, the 0.5°C rise in mean annual temperature of the northern hemisphere from the end of the nineteenth century up to the 1940s (see Sec. 3.1) extended the average growing season of plants in England by about 2 weeks. The subsequent 0.3°C decline in temperatures has been reflected in a reduction of about 10 days in the growing season. By comparison, in the coldest decades of the Little Ice Age, the average growing season in the English lowlands was shorter by 3–4 weeks in comparison with the 1930–49 average, and by a rather greater amount in the upland farms of northern England and Scotland (Lamb 1977a, p. 477).

Gloyne (1972) examined how the length of the growing season has varied during the present century at a more marginal upland site in Scotland, Eskdalemuir in Dumfriesshire (55°19′N 3°12′W, 242 m above mean sea level), and found that the growing season was generally longer over the period 1936–61 than during earlier and later periods. It also appeared that the date of commencement of the growing season (i.e. the day of the year on which the

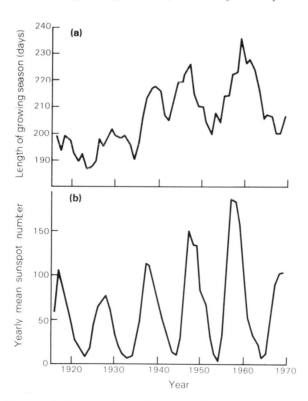

Figure 6.3 Changes in length of growing season in relation to sunspot cycle. (a) Changes in length (in days) of the growing season at Eskdalemuir, a marginal upland site in Scotland; (b) the yearly sunspot number over the same period. Redrawn from King (1973).

mean temperature rose above 42°F) showed a greater variability than the date
of termination of the growing season (defined as the day of the year on which
the mean temperature fell below 42°F). Figure 6.3 shows how 5 year running
means of the data for the length of the growing season at Eskdalemuir appear
to vary with the yearly sunspot number: the season is, on average, about 25
days longer near the sunspot maximum (King 1973) and it is suggested that the
solar cycle particularly influences temperatures in spring. It is. possible,
however, that changes in solar radiation *per se* rather than their consequences
on temperatures may have direct effects on the length of the growing season.
Thus, from 1950 to 1968 the spring flush of phytoplankton growth in the
Atlantic Ocean became progressively later, the change amounting to a total of
about 20 days over this period. This change appears to have been associated
with a decrease in the amount of incoming solar radiation: there was an
increase in cloudiness in the spring months during this period but, in addition,
a decrease in the measured strength of the solar beam may be implicated
(Lamb 1977a).

Although there are few recorded instances of changes in the altitudinal
range or northerly extent of species of wild plants as a result of climate-related
changes in the length of the growing season, historical reports (see Sec. 4.5)
attest to the considerable agricultural effects of such changes in the past.
Furthermore, if the present cooling trend persists, such effects will assume
even greater importance with a higher incidence of harvest failure in marginal
areas and reduced success of crops such as vines, peaches and maize when
grown in areas where they are not indigenous (Ford 1978c).

In addition to the effects of climate on the length of the growing season,
climatic factors also influence the amount of growth (i.e. the productivity)
achievable within a given period of time, as evidenced by the annual variations
in the yield of agricultural crops. An interesting example is that of the
productivity of the reed (*Phragmites communis*), which varies with summer
warmth; Spence (1964) has calculated that in Scotland, for example, the height
of the tallest flowering shoots in centimetres is given by

$$15 \times \text{mean temperature of the warmest month (in °F)} - 653.$$

In Norfolk during the notable summer of 1921 stems of 3.3 m were recorded,
more than 1 m higher than normal, but in the very hot summer of 1976 growth
was restricted due to low water levels. At Hickling Broad in Norfolk, there was
an average difference in height of 45 cm between compartments which had
been artificially flooded and those which had not (Hearn & Gilbert 1977). It
appears to be a sudden change in water level that is especially deleterious
(Haslam 1970, 1972). The growth of the ash tree (*Fraxinus excelsior*), which is
especially common in wetter areas, is closely correlated with rainfall during
the growing season, a relationship which shows up well in the width of the
annual tree rings (Rackham 1975). From an examination of the rings of trees
from Hayley Woods, Cambridgeshire, Rackham correlated good growth in

the years 1903, 1912, 1924, 1936, 1950 and 1958 with wet summers. Poor growth in 1915 and 1930 apparently corresponded with late summer drought in the previous years, i.e. 1914 and 1929.

As the processes of flowering and seed production occur at the end of the period of growth and development of a plant, a reduction in the length of the growing season would tend to suppress these phenomena. In addition, in many plants there appears to be an absolute requirement for high temperatures in order to initiate flowering and seed production. For example, the flowering of trees such as the beech (*Fagus sylvatica*) is very dependent on high summer temperatures: high temperatures in June or July will result in abundant flowering next spring so that mast or fruit years are correlated with the climate of the *previous* year. Thus the great quantity of fruit and seeds (although individual seed weight was reduced due to water shortage) produced by many species in England in the exceptional summer of 1976 was the result of high temperatures experienced during the summer of 1975. Some species such as the holly (*Ilex aquifolium*) always flower freely but the quantity of fruit formed varies from year to year so that good holly fruit years appear to coincide with good mast years for beech (Peterken & Lloyd 1967).

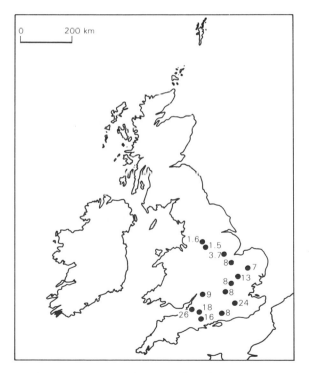

Figure 6.4 Decline in the reproductive capacity of the stemless thistle (*Cirsium acaulon*) towards the edge of its range: mean number of achenes with fully developed embryos per capitulum in September 1963. Redrawn from Pigott (1970).

Work on the stemless thistle (*Cirsium acaulon*) (Pigott 1968, 1970) has shown how the reproductive capacity of a plant can decline northwards as temperatures decrease. This thistle belongs to the family Compositae in which the flowers are aggregated into heads called capitula which bear a number of single-seeded fruits called achenes. Figure 6.4 shows how the number of fertile achenes per capitulum declines as the north-western limit of the plant's distribution in England is approached. As the rate of development increases with temperature, the first flowers of *C. acaulon* open in late June in the south but not until early August further north in Yorkshire and Derbyshire, and then it takes a further 5 weeks for the fruits to detach from the capitulum. In fact significant quantities of fertile fruits are only produced in those years when August and early September are exceptionally warm and dry. By experimental shading of the capitula, Pigott was able to demonstrate that the proportion of achenes with fully grown embryos is dependent upon the temperature experienced by the flowerheads. In a perennial plant such as *C. acaulon*, the northernmost individuals will generally be sterile except in a good year with high summer temperatures. This happened in 1976 when seed production was successful beyond the normal limits, although the resulting seedlings would not normally be able to produce viable fruits.

A number of plants are able to maintain themselves beyond the normal limits of successful reproduction by means of vegetative propagation. The dogwood (*Thelycrania* (or *Swida*) *sanguinea*), for example, fruits profusely in south-eastern England, where it is an invasive shrub, but in the north-west fertile fruit are rarely produced and it spreads almost entirely by suckers (Pigott 1970). The small-leaved lime tree (*Tilia cordata*) which, as shown in Figure 2.2, has a south-west–north-east distribution associated with the 16°C July isotherm, is rarely if ever fertile at its northern limit in Scandinavia and the Lake District of England. Laboratory experiments (Pigott 1975 and personal communication) indicate that the pollen of this species requires temperatures above 13°C for a significant proportion to germinate and above 18–20°C for rapid extension of the pollen tube. The critical event for fertilisation is pollen-tube extension. Trees in these peripheral areas must have originated in an unusually hot summer and have persisted, perhaps for centuries, owing to their ability to regenerate from suckers. An extreme example of a plant persisting in an area by vegetative reproduction when conditions are unfavourable for reproduction by seed is the case of the perennial sedge *Carex humilis*. This thermophilous plant is confined to short turf on chalk and limestone in southern England and viable fruits are extremely rare, leading Perring (1965) to conclude that summers have generally not been warm enough for reproduction by seed to occur in this species since the height of the mediaeval warm period in about the twelfth century. A similar case is that of the common salt-marsh grass (*Puccinellia maritima*; also called *Poa* or *Glyceria maritima*) which occurs in salt marshes and muddy estuaries in Europe but, in the northern part of its range, only flowers sparsely (Gray & Scott 1977). It appears to have been introduced to

Greenland by the Vikings some time in the mediaeval warm period but is now confined to the southernmost tip of the coast, south of 62°N, where it rarely flowers, and then so late in the summer that the seeds fail to ripen (Sørensen 1953).

The ability to persist vegetatively beyond the limits of reproduction by seed accounts for the apparent unresponsiveness of the distributional limits of many plants to recent and past (see the slow responsiveness of the pollen record discussed in Sec. 4.2) climatic changes. This also applies to non-flowering plants which propagate without the production of seeds. Thus, in his examination of the conditions limiting the distribution of 'Atlantic' species of bryophytes (mosses and liverworts) with a requirement for oceanic conditions, Ratcliffe (1968) noted that once such a species is successfully established in a particularly optimal year it may be able to grow under unfavourable conditions which would inhibit its establishment. Thus, although the distribution of such species is profoundly influenced by climatic conditions, they may respond only slowly to climatic changes.

Although the climatic conditions necessary for sexual reproduction in plants are generally more demanding than those for vegetative growth, the critical factor may not be solely temperature, as with the Atlantic bryophytes discussed above. For example, in the hard poa (*Catapodium rigidum*) which occurs mainly on calcareous soils in the south and east of England, drought during flowering results in the formation of abortive inflorescences. Cold springs can delay development so that flowering occurs at a time of year when susceptibility to drought is more likely (Clark 1969). In some cases reduction in flowering as a result of drought may have secondary influences on animal species which depend on plant flowers for food or oviposition sites. For example the percentage of plants of the milk parsley (*Peucedanum palustre*), an umbelliferous native of fens throughout Europe, that flower each year is dependent upon the weather. In Britain this species is the sole food plant of the swallow-tail butterfly (*Papilio machaon*), which will only lay its eggs on flowering stems protruding above the general level of the vegetation. A very low proportion of these plants flowered in the drought year of 1976 with the consequence that the breeding success of this rare butterfly was very low in that year. Similarly, flowering of the wild thyme (*Thymus drucei*) in south-western Britain was reduced in the same summer and this produced a deleterious effect on the population of the large blue butterfly (*Maculinea arion*) which oviposits solely on the flower heads of this species and which declined to such an extent that it became extinct in Britain in 1979 (see Sec. 6.7).

6.3　Growth and reproduction of animals

There has been a considerable amount of experimental work on the number of day-degrees above some developmental threshold which are required to reach a certain life history stage in poikilothermic animals, notably insects (e.g. Davidson 1944, Bursell 1964, Howe 1967, Gage, Mukerji & Randell 1976). On occasion the threshold may be equivalent to the 42°F or 5.6–6.0°C limiting temperature for the growth of plants; thus the emergence of the adult cabbage root fly (*Erioischia brassicae*) from overwintering pupae during early April to May in central England may be predicted by computing the number of day-degrees above 5.6°C (Coaker & Wright 1963). However, in other species the temperature threshold may be quite different – see for example Campbell, Frazer, Gilbert, Gutierrez and MacKauer (1974).

As with plants, the limiting effect of temperature on development and growth determines the northerly extent of many poikilothermic animals. An interesting example is the scarce aeshna dragonfly (*Aeschna mixta*) which has overwintering eggs that hatch in spring with adults emerging in the summer of the same year. Münchberg (1931) has suggested that the northerly extent of its distribution is determined by the ability of the larvae to complete development before they are overtaken by the onset of winter. In consequence this species is characteristic of the Mediterranean fauna, with immature immigrants spreading northwards. During the twentieth century climatic amelioration, however, the species colonised south-eastern England as a resident.

An important ecological result of the slowing down of developmental rate at lower temperatures is that vulnerable stages of the life history of an organism are present in the habitat for a longer period of time and so total mortality resulting from such factors as predation, parasitism and pathogen attack is greater. For example, Thomas (1976) has shown that in the black hairstreak butterfly (*Strymonidia pruni*), whose British distribution is restricted to the East Midlands forest belt (mainly Huntingdonshire, Northamptonshire and Buckinghamshire), where it feeds on blackthorn (*Prunus spinosa*), the heaviest mortality occurs in the late larval and, particularly, the pupal stages due to predation by birds, notably the willow warbler (*Phylloscopus trochilus*). The duration of these two developmental stages will be increased in cool springs (the butterfly is single-brooded, hibernating for about 9 months as an egg which hatches in March–April) and so they will be available to predators for a longer period of time. Thus, Thomas found a weak correlation ($r = 0.41$, $p = 0.02$) between crude adult counts made over 20 years at Monks Wood, Huntingdonshire, and Bernwood Forest, Buckinghamshire, each summer and the mean temperatures during the previous May and June, the time of the late larval and pupal stages. On occasions the slow rate of development resulting from low temperatures may have beneficial effects upon a population. In a number of fish species the larvae are pelagic for a period of time during which they are carried by sea currents to suitable nursery grounds (Bishai 1960). Beverton and Lee (1965) have shown how in the plaice (*Pleuronectes platessa*), which spawns in the southern North Sea, the low sea temperatures of early 1963 (about -3°C as against $+7$°C in 1962) approximately halved the rate of development. This resulted in the larvae being carried much further north-east and nearer the juvenile nursery grounds along the coasts of Holland and Germany. In consequence, the year-class of plaice was much larger than normal rather than smaller than normal as might have been expected. Examination of catch records shows that a similar occurrence happened after the 1947 cold winter. Conversely, the acceleration of the life-cycle resulting from higher than usual temperatures may produce a deleterious effect. For instance, in the very hot summer of 1976 the large blue butterfly (*Maculinea arion*), which is generally on the wing at the end of June (when the summer temperatures in that particular year were at their highest) or early July for only 3–4 days, in fact had an even shorter life-span than usual, with a consequent reduction in the number of eggs laid (see Sec. 6.7).

The effect of temperature on rate of development in poikilothermic animals may bring about changes in the number of generations produced per year. In the case of rapidly reproducing organisms, such as aphids, where population number increases exponentially, the fitting in of one or more extra generations in a climatically good year may produce a very dramatic increase in the density of the population. Thus, during the favourable weather conditions that prevailed in England in 1976 and 1977, population increases of up to ten fold per week were recorded in the grain aphid (*Sitobion avenae*) (McLean, Carter & Watt 1977). Owing to this temperature effect it is often found that the number of generations per year increases with decreasing latitude; thus the red-veined sympetrum dragonfly (*Sympetrum fonscolombii*), which is an occasional migrant to southern England where, from time to time, it manages to maintain itself for a few years (e.g. after the large migration of July 1911 breeding was maintained until the severe winter of 1916/17; Williams 1958), is univoltine (single-brooded) in Britain but bivoltine (double-brooded) in the south of France (Aguesse 1959). Even further north, some species require more than 1 year to complete a single generation: thus in Finland development of the Arran brown moth (*Erebia ligia*) lasts 2 years, with the result that in odd years the adult population may be very large but in even years the moth is very scarce (Ekholm 1975).

There are a number of examples among the Lepidoptera of species which, in Britain, usually have a single generation per year but which will produce a second generation when climatic conditions are particularly favourable. The British subspecies of swallow-tail butterfly (*Papilio machaon britannicus*) flies and lays its eggs in May and June in England and the resulting caterpillars pupate in about 6–7 weeks. In a warm dry year some butterflies will emerge from these pupae in August and produce a second generation, whilst the rest overwinter until the following May or June. A few pupae may even overwinter twice. In the long hot summer of 1959 a large second generation was produced, but most of this brood failed to survive and the population as a whole declined as a result – demonstrating again the unpredictability of biological responses to climate. The continental race, *P. m. gorganus* or *P. m. bigeneratus,* is obligately double-brooded and Bretherton (1951) suggests that it may have been established in southern England in the early nineteenth century as this subspecies is apparently illustrated by several contemporary authors. It seems to have been a regular resident for long periods, even well inland, until about the 1820s. Such records indicate that conditions must have been such as to regularly enable the second generation larvae to feed up and pupate before winter (as noted in Sec. 4.5, at the end of the eighteenth century and during the early nineteenth century summer temperatures tended to be warmer than in the present century).

The large white butterfly (*Pieris brassicae*) habitually produces two generations in Britain and in exceptional years a partial third brood occurs. In this species, as in the small white and green-veined white (*Pieris rapae* and *Pieris napi*), the two broods exhibit different colouration (see Sec. 6.6), perhaps an effect of the different temperatures that they experience. The small copper (*Lycaena phlaeas*) also usually has two generations: one flying in May and a second in July and August. The offspring of the second generation may overwinter as larvae or, in good conditions, feed up rapidly to produce a third generation flying in early October. The larvae of this generation then hibernate. The Adonis blue (*Lysandra bellargus*), which is confined to chalk and limestone areas of southern England, is also double-brooded and the obligate nature of this habit has resulted in a decline in this species since about 1950 as a result of the drop in mean temperatures which has reduced the time available for the development of the second brood. This animal, which in Britain is on the edge of its European distribution, overwinters as a first instar larva in a silken web on the undersurface of a leaf of its foodplant, the horseshoe vetch (*Hippocrepis comosa*). It starts feeding again in spring, pupates, and the first generation is on the wing in late May and early June. The eggs produced by this generation hatch quickly and the larvae develop rapidly to produce the second generation of adults which are on the wing in August and September. These lay eggs which hatch and the larvae start feeding before entering hibernation. The double-brooded habit of the Adonis blue is dependent upon sunny weather during the two periods of adult activity and the generally cool springs and poor autumns since the early 1950s have produced a simultaneous decline of the species over its range in southern England which continued at least through the 1960s.

The comma (*Polygonia c-album*) produces two generations per year, the numbers in each generation reflecting the favourableness or otherwise of the prevailing climate. Hibernating butterflies lay eggs in about April to May and these hatch to form the first generation. Some of these larvae feed up quickly and appear on the wing in late June or early July as the 'summer flight'. These individuals are morphologically distinctive as var. *hutchinsoni*. Others of this same generation grow more slowly and appear in late July and August and subsequently hibernate. The proportion of the first generation flying in June–July or July–August is largely determined through the effect of climate on developmental rate. Butterflies from the summer flight produce the second generation which flies in September as a continuation of the flight of the slow developers from the first generation. These butterflies also hibernate and pair in the spring. Very rarely, *hutchinsoni* specimens appear in the autumn and

Ford (1957) records that it is said that this happens more often in hot than cold summers. One would expect the reverse to be the case as *hutchinsoni* individuals normally result from larvae pupating in cooler conditions earlier in the year. Further investigations are necessary to clarify this point.

Data concerning the effect of climate on the number of generations per year are sparse for other insect groups in comparison with the Lepidoptera, but among the Coleoptera (beetles) it is interesting that rearing experiments on the 24-spot ladybird (*Subcoccinella 24-punctata*) (an unusual coccinellid in that it is phytophagous, being the only British member of the sub-family Epilachninae) conducted in Carlisle in the far north of England in 1922–23 (Marriner 1927) indicated that the insect produced two generations per year whilst investigations at Kew in 1974 (Richards, Pope & Eastop 1976) suggested only one. This is in accordance with our knowledge of temperature (especially the summer temperature) trends of the 1920s and the 1970s. In 1958 (Tanasijevic 1958) the species was known to produce two generations a year in Hungary, two with perhaps a partial third in Yugoslavia, and two or three in Italy.

In Section 6.2 it was pointed out that some plants are able to persist vegetatively beyond the limits of reproduction by seed. Owing to the greater mobility of animals and the comparative rarity of asexual reproduction in the animal kingdom, there are few equivalent examples from the natural fauna. However, an analogue is provided by the Portuguese oyster (*Crassostrea angulata*) which has been introduced into the oyster beds of south-eastern Britain from the Bay of Biscay, France. It is able to survive and grow in these waters but temperatures are generally not high enough for spawning to occur.

In those lower animals (including protozoans, coelenterates, rotifers and triclads) which do exhibit asexual reproduction, unfavourable climatic conditions often induce the onset of sexual reproduction. This contrasts with the situation in plants, but is clearly adaptive in that some form of resting stage is usually produced which enables the animal either to survive the unfavourable conditions or else disperse to a more favourable habitat. A similar phenomenon perhaps occurs in plants (including trees, such as the apple – see Landsberg 1979), where a reduction in the water supply is frequently associated with flower bud formation. This would appear to be a mechanism ensuring the production of propagules when the survival of the plant is endangered. Among the more advanced animal groups, asexual reproduction is an important feature of the life-cycle of aphids, which are parthenogenetic (i.e. develop from unfertilised eggs) and viviparous (i.e. the embryos develop inside the female and the young are born live). In most species, however, sexual morphs are produced in the autumn (in response to factors such as changing day-length and food quality as well as temperature) and then the oviparous females lay overwintering eggs. Some species have forms which exist parthenogenetically all through the year (anholocyclic forms) together with other forms which lay overwintering eggs in autumn (holocyclic forms). The parthenogenetic forms can survive mild winters but are killed by severe winters after which anholocyclic clones develop again from the holocyclic survivors. The bird cherry–oat aphid (*Rhopalosiphum padi*) (so-called because it spends from autumn to spring on the bird cherry tree, *Prunus padus,* but in summer, when the bird cherry leaves are mature and provide a poor source of food, it migrates to grass species) is an interesting example as it is holocyclic in north-western Britain, but in the south-east also exists as an anholocyclic strain.

There are also some analogues in the animal kingdom of the absolute requirement in certain plants for a period of high temperature in order to initiate sexual reproduction. The animal examples are found particularly in the cold-blooded vertebrates; thus Volsøe (1944) reports that after arousal from their winter hibernation, male adders or vipers (*Vipera berus*) undergo a period of basking in the sunshine for about 4 weeks before mating, during which time spermatogenesis occurs. The eggs of the adder develop inside the oviduct of the female so that the young are born fully developed, a phenomenon comparable with the viviparity displayed by mammals and hence termed ovo-viviparity. In the extreme north of Europe (e.g. Finland; Vaino 1932), however, the short duration of the summer retards the development of the eggs with the result that successful breeding only occurs every second year. The common lizard (*Lacerta vivipara*) is also viviparous, the only member of its genus to display this characteristic, but there are some rather contentious reports that it may be oviparous (i.e. egg-laying) in certain locations in southern Europe (e.g. Lantz 1927, disputed by Panigel 1956).

6.4 Development of immature organisms

It has been pointed out in the previous two sections that the climatic conditions necessary for the sexual reproduction of plants and, to a lesser extent, animals are generally more demanding than those necessary for non-reproductive growth. In addition, the establishment and development of immature plants and animals frequently requires more favourable climatic conditions than those necessary for the growth of mature organisms, and any extreme of climate frequently has a disproportionate effect on the earlier stages of the life-cycle. Thus, the vulnerability of seedlings to drought in spring and early summer, before they have developed an adequate root system, is well established. A notable example is the aspen tree (*Populus tremula*), the seeds of which require moisture within 1 week of seedfall if they are to germinate and continuous moisture throughout the early stages of growth. Seedling mortality during the first summer is very high in this species so that, although seeds are generally produced every year, recruitment to the adult population is an extremely rare occurrence. Once established, however, the aspen is able to withstand summer droughts. Adequate moisture is clearly important in the early development of those animals with an aquatic life-stage (e.g. amphibians, dragonflies) as well as terrestrial species whose eggs are subject to water loss. Pollard (1975) reports that drying of the soil surrounding the egg-cavity resulting in desiccation is a major cause of egg mortality in the edible or Roman snail (*Helix pomatia*). Almost always some eggs in a batch fail to hatch for this reason and sometimes all the eggs dry up. Owing to the calcium requirements of adult snails with their calcareous shells, populations are restricted to chalk or limestone soils which, as they are free-draining, tend to be drier. They are also often located in areas of relatively low rainfall. There is thus a conflict between the soil requirements of the adults and those of the eggs. Furthermore, adult snails prefer south-facing slopes with high insolation (see Sec. 5.2); a preference which is also in conflict with the requirements of the eggs as the higher temperatures experienced on such slopes tend to dry out the soil.

The immature stages of organisms are more susceptible to extremes of cold and late frosts often kill or injure young plants and animals whilst more mature organisms can survive such events. In some cases younger organisms are more vulnerable to cold winters than are adults: in the winter of 1962/3 the better survival of second winter cockles (*Cardium edule*) at Shoeburyness, Essex, was shown by their changed proportion from 33% of the total population in November 1962 to 90% in February 1963 (Crisp *et al.* 1964). Climatic extremes may also result in developmental abnormalities. For example after the 1962/3 winter the sea anemones *Anemonia sulcata* around Plymouth were found to have abnormally short tentacles. Similarly, cold springs result in a high incidence of abnormalities in tadpoles of the common frog (*Rana temporaria*), which breeds earlier than other amphibians such as the natterjack toad (*Bufo calamita*) and the newts (*Triturus* spp.), which are in consequence more vulnerable to the effects of summer drought.

6.5 Phenology

The effect of climate upon the metabolism of organisms discussed in earlier sections of this chapter, and especially its influence on the rate of growth and development of animals and plants, produces variations in the seasonal timing of biological events which comprise the objects of study of the subject of phenology (see Leith 1974). Long series of phenological records have provided useful evidence concerning past climatic events in the historical period. The earliest phenological observations were the 'Wu Hou' observations of the Chinese begun in the Chou (1027–221 BC) and the Ch'in (221–206 BC) dynasties in order to effect the correct timing of agricultural activities. These were initially an annual series of twenty-four fortnightly records, but later on the year was divided into seventy-two 5 day periods (Chu 1963). Many of these Wu Hou were incorporated into the poems of the T'ang (AD 618–907) and Sung (AD 960–1280) dynasties and so have been preserved. In Kyoto (35°N 136°E), the capital of Japan until 1868, the dates of the blossoming of the cherry trees (probably *Prunus yedoensis*) have been kept intermittently since AD 812 (Arakawa 1955) and form the longest known record of a phenological event. These dates are a function of the integrated spring temperatures and have yielded valuable information on the changing climate (Yoshino 1974). There are currently 1500 phenological observing posts maintained throughout Japan (Chu 1963). In Europe, records of the dates of harvests in France, Germany, Switzerland and Luxembourg going back to the fourteenth century constitute phenological data from which useful climatic information concerning spring and summer temperatures can be extracted (Müller 1953, Le Roy Ladurie 1972). The first regular network of phenological observing posts in Europe was that organised in Sweden by the great biologist Linnaeus (1707–78). This consisted of eighteen stations but was only in operation from 1750 to 1752. In 1755 Alexander Berger compiled a calendar of the year in Sweden and in 1761 Benjamin Stillingfleet published an English translation of this as *The calendar of the flora,* together with some observations that he had made when staying at Stratton Strawless with Robert Marsham in 1755. This publication inspired a number of eighteenth century naturalists, including Gilbert White (1720–93) whose observations cover the years 1768–93, to start keeping phenological records. White's *Naturalist's calendar* was published posthumously in 1795.

The Marsham Phenological Record (Marsham 1789, Southwell 1875, 1901, Margary 1926) constitutes the longest such series known from England. Robert Marsham FRS (1708–97) began to compile annual tables of what he called the 'Indications of the Spring' from 1736 at his estate, Stratton Strawless Hall, Norfolk (52°44′N ,1°16′E), about 7 miles north of the city of Norwich and 12 miles south-west of the North Sea coast. The record was continued, with occasional gaps, by his descendants until 1947, although with a slight change of location in 1858 when the family moved $1\frac{1}{2}$ miles further north to Rippon Hall (52°45′N 1°16′E). The record comprises twenty-seven phenological events: four flowering events (snowdrop, wood anemone, hawthorn and turnip), thirteen leafing events (hawthorn, sycamore, birch, common elm, mountain ash, oak, beech, horse chestnut, chestnut, hornbeam, ash, lime and maple) and ten events concerning animals (the first appearance of the swallow, cuckoo and nightingale; the first calls of the night jar, thrush and ringdove (probably any pigeon); the first signs of nest-building by rooks and the appearance of young rooks; the croaking of frogs and toads and the appearance of the brimstone butterfly). The date of flowering was determined when the centre of a blossom was visible and leafing was taken to be the date when the leaves were of recognisable shape. In some respects the recording methods are unsatisfactory; for instance, the *first* plant or tree *anywhere* on the estate to flower or leaf was noted, but Margary (1926) showed that there was very good correspondence (correlation coefficients between 0.81 and 0.93) between the Marsham Record from 1891 onwards and the more rigorous annual Phenological Reports of the Royal Meteorological Society described below.

Margary also showed that the annual means of seven Marsham events (flowering of snowdrops and hawthorn; leafing of birch, oak, beech, horse chestnut and lime) varied in close agreement with the mean January–May temperature (Margary used the only available temperature series, that of Glaisher (1849, 1850) for Greenwich, going back to 1771). This is demonstrated in Figure 6.5 which indicates an advance of about 5 days for each 2°F rise in mean temperature. The dates of bird arrivals were found to bear little relationship to the temperature values, although the cuckoo dates showed some agreement with April temperature. Kington (1974) examined changes in the dates of leafing of six indigenous or long-naturalised trees in the Marsham Record (hawthorn, birch, horse chestnut, lime, beech and oak) from

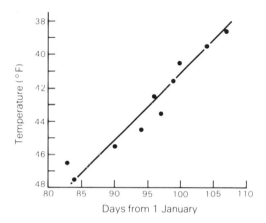

Figure 6.5   Marsham phenological record: relationship between mean January–May temperature and annual means of seven flowering and leafing events. Redrawn from Margary (1926).

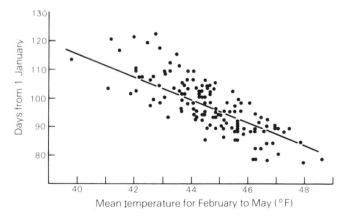

Figure 6.6   Marsham phenological record: relationship between the average date of six leafing events and the mean temperature of the 4 month period February–May. Redrawn from Kington (1974).

1745 to 1925 and was able to identify periods with a preponderance of late or early seasons. Using different groupings of monthly temperatures from the Manley central England series (Manley 1953 – see Section 4.5) he found the maximum correlation coefficient (-0.81) between the mean date of the six leafing events and the mean temperature of the four-month period February to May (see Fig. 6.6). This relationship indicated that the mean date was advanced by 4 days for every $1°F$ rise in mean temperature.

The origin of the Royal Meteorological Society's Phenological Reports (Clark 1936, Craddock 1974), mentioned above, was due to the initiative of Rev. Thomas A. Preston, a master of Marlborough College and organiser of the Marlborough College Natural History Society. Preston edited the RMS Phenological Reports from 1875, when there were seventeen stations, until 1888, when there were ninety-nine. His suggested list of seventy-one plant species, eighteen birds and eight insects for observation proved too taxing for many of the reporters and in 1891 his successor as editor, Edward Mawley, reduced the list to thirteen plants, six birds and five insects. In 1914 a further twenty-òne migrant birds were added with the ending of the reports collated by the British Ornithological Union. The last published Phenological Report was that for 1947, by which time there were over 500 stations reporting. The reasons for the ending of the Reports are explained by Hawke (1953) thus:

> . . . a suggestion came before Council that an association of meteorologists was not the best possible authority to conduct a modern phenological survey. This view having been accepted, it was decided that the Phenological Report for 1947 should be the last to appear under the Society's auspices. Collection of the observations was continued for a few years in the hope that some biologically qualified body might take over their analysis and publication, but this did not happen, and in 1951 the organisation was disbanded.

However, from quite early on, there was some attempt at analysis of the data being collected. Annual floral isophenes (lines of equal date of flowering) using 10 day intervals were produced from 1916, although the time-interval was changed to 7 days in 1918; and in 1922 a chart of migrant isophenes (lines of equal date of arrival of migrant species) was also introduced (Clark & Margary 1930). In 1926 average isophenes for the twelve spring- and summer-flowering plants (hazel (*Corylus avellana*), coltsfoot (*Tussilago farfara*), wood anemone (*Anemone nemorosa*), blackthorn (*Prunus spinosa*), garlic hedge mustard (*Sisymbrium allaria*), horse chestnut (*Aesculus hippocastaneum*), hawthorn (*Crataegus oxyacantha*), white ox-eye (*Chrysanthemum leucanthemum*), dog rose (*Rosa canina*), black knapweed (*Centaurea nigra*), harebell (*Campanula rotundifolia*) and greater bindweed *(Convolvulus sepium)* – the thirteenth species being the autumn-flowering ivy (*Hedera helix*)) for the previous 35 years were published (Fig. 6.7). For each year from 1891 onwards, maps showing lines of equal unseasonableness (called 'isakairs') at 5 day

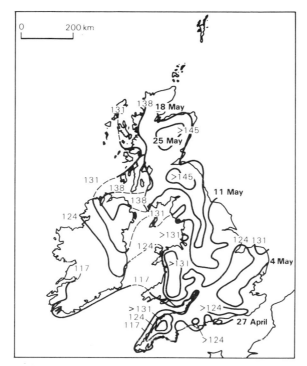

Figure 6.7 Floral isophenes: averages from the Royal Meteorological Society's Phenological Reports for the 35 years 1891–1925. Redrawn from Clark and Margary (1930).

intervals before or after the average date of the region were prepared, based on the 35-year average floral isophene map (Clark & Margary 1930). Unlike the isophene maps, isakair maps eliminate the effects of features such as latitude, altitude, soil conditions, distance from the sea, etc., which are constant from year to year, and bring out the annual variations in climate.

Jeffree (1960) examined the data for eleven flowering species recorded consistently from 1891 to 1948 and produced 58-year means of their flowering times for Britain as a whole (see Table 6.1) weighted according to the actual number of observers in each of the eleven meteorological districts into which the country was divided. Jeffree also calculated the means for each of the eleven meteorological districts separately and found the dates for the latest and earliest districts varied between 10 days for harebell and 28 days for ivy with a mean of 21 days. Thus, over a distance of about 400 miles from southern England to roughly Aberdeen, the mean flowering time becomes 21 days later, i.e. 1 day for every 19 miles. This works out at 3.7 days per degree of latitude, which corresponds quite well with Hopkin's 'bioclimatic law' calculated for the eastern USA, which states that the indicators of spring such as leafing and flowering occur 4 days later for every 1° latitude northward, 5° longitude westward or 400 feet of altitude.

Table 6.1   Fifty-eight-year (1891-1948) weighted means for flowering times in Britain of eleven plant species recorded in the Royal Meteorological Society's Phenological Reports (from Jeffree 1960). Day 1 = 1 January, etc.

Plant Species	Day of year
hazel	41.3
coltsfoot	62.4
wood anemone	88.3
garlic hedge mustard	113.7
horse chestnut	128.5
hawthorn	133.2
white ox-eye	149.5
dog rose	159.3
greater bindweed	190.6
harebell	190.8
ivy	270.8

In conclusion, it is quite clear that defining the seasons in terms of calendar dates is not biologically meaningful: 'biological spring' may vary by several weeks from year to year. As discussed in earlier sections of this chapter, it is possible to define the start of the growing season of plants in terms of a temperature threshold, but animals and plants are affected by a number of climatic variables other than temperature. The advantage of studying phenology is that the dates of biological events provide an integration of all the diverse climatic influences that affect living things. However, one must bear in mind that in some organisms the timing of events will be controlled by such factors as changing day-length, which does not vary from year to year, and so such species will apparently not respond to annual differences in climate.

In any phenological study most information will be obtainable from plants which, being immobile, are affected by the weather at a single known site. In contrast the arrival of migratory birds will provide equivocal information as it will be partly determined by the conditions at their overwintering home as well as *en route*. Whatever organisms are being studied, one must take into account genetically determined individual variability: some individuals of a species are habitually early and others are always later. This is a problem in interpreting the Marsham Record where the first occurrence of an event anywhere on the estate was noted. The early RMS Phenological Reports similarly ignored this point until 1891, when Mawley suggested that, instead of recording the flower first seen, each report should be from one chosen plant or group of average exposure. A related problem is that some species comprise a number of distinct subspecies which may have differing phenological characteristics. This was found to be the case with the black or lesser knapweed (*Centaurea nigra*) which was one of the thirteen plants on the RMS list and proved to be extremely variable in its date of flowering (Turrill 1937). There are two distinct subspecies of this plant in Britain, as well as intermediates, and it was not possible to distinguish which species was being referred to after a record had been sent in. Even separating distinct species such as the two main oak species *Quercus robur* and *Q. petraea* and the three elms *Ulmus glabra, U. procera* and *U. carpinifolia* may present difficulties for the non-botanist. Another factor to be borne in mind is that the date of a phenological event may not be determined by the conditions of the immediate past (e.g. spring temperatures) but may be related to conditions in the previous season. In addition other factors, such as competition, may lead to an earlier or later date than would be optimal on purely climatic grounds. Lastly, the effectiveness of any phenological survey is determined by the diligence of the reporters. Sparsely populated areas will always be under-reported and some reporters may return incomplete records. Thus, in 1926 (Clark & Margary 1930), out of over 300 returns of flowering events only some forty observed all thirteen flowers. A further complication that the RMS found was that there was a turnover of reporters of the order of 15–20% *per annum*. However, despite all these difficulties it is clear that careful recording of a number of phenological events (preferably those relating to perennial plants such as trees where the same individual could be observed in successive years) can provide a valuable indicator of the onset of spring in biological terms which would also have useful agricultural application.

6.6   Animal colouration

As discussed in Section 2.2, darker coloured insects absorb 20–50% more solar radiation than light ones (Digby 1955) and this effect will clearly be more important in colder climatic conditions. Watt (1968) investigated the fifteen American *Colias* butterfly species (Lepidoptera: Pieridae) and found that, both within and between species, melanism (genetically determined dark colouration) increased at higher altitudes and latitudes. Similarly, the *Erebia* species (Satyridae) of the Arctic and alpine areas are frequently almost black. Clench (1966) pointed out that heat exchange in butterfly wings is most effective in the distal parts where the veins are narrower with a greater total surface area, and in line with this many butterflies have a dark border to their wings. Clench also indicated that often the female of a species is thermally darker than the male and this may be associated with the higher metabolic demands of egg maturation or else the fact that the necessity to oviposit may force the female to be active in unfavourable weather or take her into shaded microhabitats.

   It is clear that the regional variation in the abundance of dark or melanic forms of a number of Lepidoptera such as the peppered moth (*Biston betularia*) (Kettlewell 1958) results from the selective predation by birds of light-coloured forms in areas of high industrial pollution (Kettlewell 1961, 1973). However, the association of melanic forms of the two-spot ladybird (*Adalia bipunctata*) (Coleoptera) with atmospheric pollution can not be explained by the same mechanism as ladybirds are distasteful to birds and exhibit warning colouration. Instead, it has been suggested that some component of the polluted atmosphere is less toxic to the melanic morph, which thus predominates in smoky environments (Creed 1971). An alternative explanation is that of Lusis (1961) who suggested that, in Europe, melanic forms are most common in places with a maritime climate (i.e. cloudy conditions) and industrial centres because of their increased ability to utilise solar radiation for raising their body temperature. Benham, Lonsdale and Muggleton (1974) found a highly significant negative correlation ($r = -0.59, p < 0.001$) between the annual total of the number of hours of bright sunshine and the frequence of melanic morphs of *A. bipunctata* at 151 sites in Britain and experimental illumination by a 40 W bulb at a distance of 10 cm resulted in melanic morphs attaining an internal temperature 1°C higher than typical forms after about 10 min (Muggleton, Lonsdale & Benham 1975). The case of *A. bipunctata* shows that the possibility of a thermal advantage should not be ignored when evaluating the incidence of melanism, particularly in the current situation of a decreasing temperature trend. This thermal advantage or 'energy subsidy' (Odum 1967) could be quantified in metabolic terms by means of respirometry studies of melanic and non-melanic forms of a given species.

One clear case of the influence of climate on the development of melanic forms is that of the sycamore aphid (*Drepanosiphum platanoides*), which hatches in March and is present until early November. In this species, aphids present in spring and in autumn develop melanic pigmentation in response to the low temperatures that they experience (Dixon 1972) and it is during these seasons that reproduction and growth are maximal. In summer reproduction declines as the nitrogen content of the phloem of the host tree is reduced and so the nutritive quality of the aphid's diet declines.

Another example of a genetic polymorphism where selection may be influenced by climate relates to differences in shell colour and banding morph frequencies in snails of the genus *Cepaea*. *Cepaea hortensis* tends to prefer cooler, damper habitats than *C. nemoralis*, which is favoured by warm dry conditions (Harvey 1974, Cameron 1970a, b), and it seems that morph frequencies in the former species are maintained by climatic selection: the bandless morph being most frequent in cold, humid areas while the band-fusion morph being most frequent in temperature-stable areas (Arnason & Grant 1976). In the case of *C. nemoralis*, the snail tends to be rare where temperature variation is great and there is evidence (Richardson 1974) that such variation affects morph frequencies with minimum night temperature probably being the limiting factor (Bantock & Price 1975). In particular the brown morph (pink and yellow being the more common forms) is largely restricted to the northern part of the species' range (Cain & Currey 1963) and appears to be more common on north-facing slopes and in valley bottoms likely to accumulate cool night air (Bantock 1974, 1980).

In the double-brooded large white butterfly (*Pieris brassicae*) (as well as the small white (*P. rapae*) and the green-veined white (*P. napae*)) the spring brood, active in April–June, and the summer brood, active in July–September, exhibit different colouration, presumably climatically induced. Larvae which pupate in autumn will be subject to cooler temperatures than those which pupate in the middle of the summer and give rise to the second generation. The spring brood (originally considered a distinct species, *P. chariclea*) generally has greying tips to the forewings, whilst in the summer individuals the tips are usually black (Howarth 1923): this may present a thermal advantage to the second generation according to the views of Clench outlined above. An analogous effect of cooler temperatures being associated with paler colouration (although note that in the Bath white, *Pontia daplidice*, the first brood is smaller and *darker* than the summer brood) is found in the Camberwell beauty (*Nymphalis antiopa*) where it was formerly thought that the British-bred individuals had a white border to the wings whilst continental specimens had a cream border. In fact, the colouration of the border is connected with the latitude of the breeding area as specimens from Scandinavia have a paler border than those from southern Europe. The prevalence of specimens with pale borders in Britain indicates their Scandinavian provenance. Howarth (1973) quotes E. A. Cockayne's view that border colouration in the Camberwell beauty is associated with a scale defect which is more prevalent in Scandinavian specimens. He also describes the work of Standfuss, who clearly elucidated the effect of temperature in producing another type of variation in the species, namely variation in the width of the border, which may be so broad as to partly or completely hide the blue spots which are usually visible on the dark band just inside the border (ab. *hygiaea* Heydeni or ab. *lintneri* Fitch). This form has been produced experimentally by placing pupae in a temperature of 110°F (43°C) for a 1 h period four times a day for 3–4 days. Thus, wild-produced specimens of this type presumably result from pupae which have by chance been subjected to short periods of high temperature.

A very indirect effect of climate upon animal colouration is suggested by the report by Richards and Waloff (1954) that the green varieties of several species of grasshopper (Acrididae) became scarce after the very dry summer of 1949 in England. This indicates that green forms suffered heavier predation owing to the fact that they were conspicuous against the predominantly yellow-brown vegetation resulting from the lack of rainfall.

Although the examples so far discussed in this section relate to the influence of climate on the colouration of poikilothermic animals, the coat colour of homeotherms is also adapted to prevailing climatic conditions and is of particular significance in extreme environments. The association with the influence of climate may be direct as in the case of the predominant black coat of desert goats which, by absorbing more of the sun's energy on cold days, increases the animal's body temperature and significantly decreases its oxygen consumption (Dmi'el, Prevulotsky & Shkolnik 1980). Or it may be indirect, as in the seasonal change in colouration undergone by species such as the ptarmigan (*Lagopus mutus*) and the mountain hare (*Lepus timidus*) in order to maintain their inconspicuousness in a habitat which is snow-covered for part of the year. In the latter species Jackes and Waston (1975) found that although colour change from brown to white in early October in Scotland was initiated by decreasing day-length, its progress in the following 10–12 weeks was related to air temperatures. From December to February snow-lie was the controlling factor and then the change from white to brown was again controlled by day-length, but with some influence of temperature accounting for the observation (Watson 1963) that animals regain their brown colour more slowly in cold snowy springs than in mild ones.

6.7 Climate and activity

The dependence of locomotory activity upon temperature is well established for poikilotherms (Fry 1947, Mellanby 1939), particularly flying insects (Williams 1940, 1961, Southwood 1960, Taylor 1963) but also for epigeic (species active on the surface of the ground) invertebrates such as beetles (Greenslade 1964) and spiders (Ford 1978a), and even marine species such as the razor shells that are reported (Crisp *et al.* 1964) to have completely lost their ability to burrow in the low temperatures of the severe winter of 1962/3 and so were left exposed on the surface. The phenomenon similarly applies to poikilothermic vertebrates, especially reptiles where for each species a 'critical minimum temperature' (Spellerberg 1973, 1977), the lower body temperature at which locomotory ability is disorganised and the animal loses its ability to escape from conditions that may lead to its death, can be determined and appears to have ecological significance. Other climatic factors such as windiness, sunshine and rainfall also influence the activity of animals and so may reduce the time available for such biological necessities as obtaining food or a mate or laying eggs.

In addition, reduced locomotory activity may increase the susceptibility of an organism to predation, rather like the effect of a slowing down of developmental rate (Sec. 6.3). For example, usually the first flight after emergence in the emperor dragonfly (*Anax imperator*) is closely synchronised with 95% of adults departing within the first 30 min of dawn, i.e. before the first birds begin to feed. If low temperature postpones the flight, blackbirds (*Turdus merula*) have the opportunity of pecking dragonflies off their emergence supports and this may be a significant mortality factor (Corbet 1957). As temperature increases around midday, the activity of dragonflies is generally reduced so that many, such as *Aeshna cyanea*, exhibit a bimodal activity pattern with peaks in the morning and afternoon (Moore 1953). In the southern part of its range in France, the activity peaks get further apart so that this species tends to fly only at dusk and dawn when the weather is hot (Robert 1958).

The activity of butterflies is very dependent upon the weather and this must be borne in mind when attempting to assess their abundance visually, e.g. by a transect count (see, for example, Pollard *et al.* 1975). Douwes (1976) studied the effects of weather variables on the activity of the scarce or middle copper (*Lycaena* (*Heodes*) *viriaureae*) in Sweden. This species is common in woodlands in northern Europe, where it flies from early July to late August, and it may have occurred formerly in Britain, probably becoming extinct in the early eighteenth century. The butterfly's activity is related to air temperature, solar radiation and wind velocity. It is inactive in cool, cloudy or windy conditions but quickly responds to short periods of sunshine by an immediate increase in activity. Many butterflies are heliotherms (see Sec. 2.2) and bask in the sun in order to raise their body temperature to a level that permits activity even when air temperatures are low (Clench 1966). Douwes (1976) found that in full sunshine the body temperature of the scarce copper rose 10–15°C above ambient. In contrast, moths, many of which are nocturnal, are myotherms and use muscular work to generate heat. This may be the result of normal activity, though many moths use high frequency, low amplitude wing vibration as a means of warming up (Dorsett 1962). This type of muscular pre-flight warm-up is also found in the Coleoptera and Hymenoptera (Krogh & Zeuthen 1941, Heinrich 1972, 1974).

In the small copper (*Lycaena phlaeas*), Dempster (1971) found that sunshine was a controlling factor for oviposition: whenever a cloud obscured the Sun, the female's activity ceased until the Sun re-emerged. Dempster considered that the low sunshine figures for June, July and August in the early 1960s probably explained the decline in the species over this period; a decline that was halted in 1969 when sunshine figures were above the 30-year mean. Similarly, in the case of the swallow-tail butterfly, the British race of which is now confined to a number of sites in the Norfolk Broads, the number of eggs laid each year varies with the weather at oviposition (Dempster, King & Lakhani 1976); thus the warm sunny June in 1973 produced the highest density of eggs recorded. On occasion a high proportion of eggs laid are infertile and this is a consequence of failure to mate as a result of bad weather during the flight period.

The flight period of some butterflies is extremely short and poor weather during this period, especially a climatic trend resulting in a number of poor seasons, e.g. a series of cold springs or wet summers, may have serious results on the viability of a population, as was the case with the large blue butterfly (*Maculinea arion*) in Great Britain. In the nineteenth century this species was quite widely recorded in most southern counties of England, but later in that century its distribution had become very localised, although it was still quite common in Northamptonshire, Gloucestershire and south Devon. In the 1880s it was thought to be on the verge of extinction due to adverse weather conditions until new populations were discovered in the Atlantic coastal area of north-eastern Cornwall and north-western Devon. With the climatic amelioration of the early part of the present century numbers increased and many new sites were discovered. However, by the 1930s it was again on the decline, by 1960 it was only found in Cornwall, Devon, Gloucestershire and

Somerset, and by 1973 was restricted to a single site in Devon where it apparently became extinct in 1979.

Over-collection of this large and attractive butterfly (associated with the impossibility of artificial rearing due to the complexity of its life-cycle, as described below) was certainly a contributory factor in its decline, but its habitat requirements are very demanding. It requires the presence of wild thyme (*Thymus drucei*) for food and oviposition and as this plant is characteristic of early successional stages it tends to be crowded out by scrub encroachment, a process which has increased with the demise of the rabbit (*Oryctolagus cuniculus*) population due to the outbreak of myxomatosis in 1954. Benham (1973) has suggested that the heyday of this butterfly was soon after the recession of the glaciers when wild thyme would have been a significant component of the flora colonising the bare ground left behind in the wake of the retreating ice. It probably only survived in very limited localities until the clearance of the forests and the introduction of the rabbit in the late twelfth century when its food plant would have become abundant again. More recently the plant has declined with the expansion of agriculture and the regular burning of the hillsides in winter, known as swaling, to improve grazing. Thyme prefers well drained sites and frequently occurs on the mounds produced by the yellow ant (*Lasius flavus*). However, the butterfly requires the presence of red ants of the genus *Myrmica* in order to complete its life-cycle. The species *M. rubra* and *M. ruginodis* generally occur in wetter habitats, so the species of main concern are *M. scabrinodis* and, most importantly, *M. sabuleti*, the latter flourishing only under heavy grazing which maintains a very short turf. Ants of this species take fourth instar larvae of the large blue into their nests in autumn and tend them for the sake of nutritious secretions produced by the honey gland situated on the seventh abdominal segment of the caterpillar. The butterfly larvae feed on the ant larvae before hibernating in the autumn and again in the spring. In late May they pupate and emerge about 3 weeks later.

This is the point at which climate becomes important because the adult has a very brief life and, if the weather is unfavourable during the short period in late June to early July that it is on the wing, its activity will be reduced so that it does not succeed in mating and laying eggs. From mark–release–recapture experiments in 1964, when the weather was favourable, Hunt (1965) found that the average life of a large blue butterfly was 3–4 days with a maximum of 9 days (see Sec. 6.3). Spooner (1963) reviews the evidence of the rises and falls of the large blue population over the past 100 years and shows that increases are associated with climatically good years and decreases with bad years. The general downward trend over past decades has primarily been due to loss of suitable habitat. This brought the population to a low level at which climatic effects became critical and the trend to increased variability of climate in the 1970s discussed in Section 3.2 eventually brought about the demise of the species. Cold wet weather in the flight season such as that experienced in 1974 has always produced, via a reduction in the number of eggs laid, a decline in the population in the subsequent year. However, the anomalous summers of 1975 and 1976 with high temperatures and excessive drought which reduced the flowering of the thyme plants produced a further decline so that in 1977 only sixteen adults emerged at the last remaining site. By this time numbers had fallen so low that the non-synchronous emergence of the few adults produced in that year, as well as the effects of the poor summer, meant that few butterflies were able to mate. In consequence only five adults were counted in 1978; twenty-two emerged in 1979 but no viable eggs were produced, hence it was concluded that the species was almost certainly extinct in Britain.

Muggleton (1974) suggested that *M. arion* may have adjusted the date of its emergence in line with earlier climatic changes, at least in Gloucestershire where the species appeared slightly earlier than further south. There is an almost complete record of the dates of appearance of adults of *M. arion* in this county from 1858 to its disappearance in 1960. If the data are split up into 25-year periods, it is found that the peak of first appearances occurred in the third week of June between 1851–75; during 1876–1900 and 1901–25 the peak occurred in the fourth week of June and between 1926 and 1950 the peak returned to the third week. Muggleton writes that this 'suggests an ability by *arion* to adjust its life cycle to climatic changes', but it is more likely to be nothing more than a passive phenological effect of temperature on developmental rate such as described in Section 6.3.

Although, as in the case of the large blue, rainfall generally reduces the activity of insects, it facilitates the activity of snails such as *Helix pomatia*. These animals hibernate between September and April and the precise timing of arousal is dependent upon mild, wet weather. Subsequently they may become completely inactive again for long periods in dry weather. As activity is a prerequisite, mating is only initiated in wet weather, mainly in May and June. Wells (1943) suggested that activity in *H. pomatia* is initiated by the physical stimulation of raindrops falling on the shell but Pollard (1975) has shown that there may also be a period of early morning activity solely in response to dew. The activity of slugs is generally controlled by temperature, together with vapour pressure deficit which affects the rate of evaporation of water from the animal's surface (Webley 1964, Crawford-Sidebotham 1972).

7
Transport effects of climate

The previous chapter described some of the direct effects of changing climatic factors such as temperature and insolation on living things. However, as described in Section 3.2, climatic changes are fundamentally associated with changes in the atmospheric circulation, and especially in the strength of the upper westerlies and the position of their associated ridges and troughs. Such changes are accompanied by variations in the trajectory, frequency and strength of wind and sea currents and these variations may produce changes in the range of those species which are susceptible to transport by such currents.

7.1 Ocean current-assisted transport

Marine algal fragments may be transported considerable distances by water currents and this accounts, for example, for the occasional occurrence of viable portions of the West Indian *Sargassum* on the eastern shores of the Atlantic, e.g. along the coast of Devon and Cornwall. Similarly, a small beetle of the genus *Microlymma*, which occurs amongst seaweed on the eastern coast of the USA, is sometimes carried out to sea on rafts of floating vegetation and by this means has spread around the shores of northern Europe. Dixon (1965) writes that viable specimens of *Gelidium cartilagineum* have been found as far north as Shetland, Heligoland and Schleswig-Holstein, although the northern limit of distribution of this species is generally given as the Canary Islands. The alga *Colpomenia peregrina* has ballon-like thalli which enable it to be transported by the prevailing currents and so colonise north-western France and southern England in the early years of this century (Dixon 1965). It gradually spread around the British Isles and reached Denmark and southern Norway in the mid-1950s. In stormy conditions, algae may be transported to new localities attached to pebbles and Lucas (1950) describes large scale transport of algae from the southern entrance of the English Channel north to the coast of Holland on floating objects such as timber. Such naturally occurring transport is clearly related to the strength and direction of the prevailing ocean currents, but the situation is complicated by the suggestion (De Valera 1942, Powell 1957) that attachment to shipping may be responsible for the transport of some marine algae. The giant Pacific alga *Sargassum muticum* ('Japweed'), which was first noted at Bembridge on the Isle of Wight in February 1973 and subsequently spread to the Hampshire coast, may have arrived in this fashion.

It is frequently found that plants do not reproduce sexually at the northern limits of their distribution (see Sec. 6.2). This is the case in many members of the algal group Rhodophyta (red seaweeds) where it is apparent that spores can be transported to the north of the limit for the expression of the species' reproductive potential yet still germinate and develop normally (Dixon 1965). The resulting plants are sterile but persist by vegetative reproduction. Although the disseminules (portions of a plant that are dispersed) of most land plants are killed by immersion in sea water, the fruit and seeds of a number of littoral species are dispersed by sea currents, the most familiar example being the coconut *Cocos nucifera*. Examples from around the coasts of Britain include the sea sandwort (*Arenaria peploides*) and the oyster plant (*Mertensia maritima*), which has been retreating northwards in recent decades.

Many marine animals, if not planktonic as adults, at least have immature stages which are planktonic and thus susceptible to the influence of ocean currents. A celebrated example being the leptocephali larvae of the eel (*Anguilla anguilla*) which are born in the Sargasso Sea in the western North Atlantic and then drift across to Europe in the Gulf Stream. This journey takes about 3 years, during which time the larvae experience an exceptionally high (about 99%) mortality. However, in considering the effect of climatic changes on the distribution and abundance of marine organisms it is very difficult to distinguish transport effects associated with changes in the strength and direction of ocean currents from the direct effect of a change in sea temperature on the range of a species. To some extent the two effects can be differentiated in the case of the cod (*Gadus morhua*) whose northern limit of distribution is determined by the location of the 2°C isotherm of sea surface temperature. The animal's physiology, especially its osmoregulation (regulation of the body's water balance), is disturbed below this temperature (Woodhead & Woodhead 1959), which is roughly equivalent to the transition between waters of Atlantic and Arctic origin (Beverton & Lee 1965). Thus Templeman (1965) reports mass mortalities of cod (and other fish) which had become trapped in a mass of cold Arctic water off Newfoundland. Temperatures between 4 and 7°C seem optimal for reproduction and early survival and southerly cod populations show a negative correlation between temperature and year-class numbers (Martin & Kohler 1965).

Prior to the early part of the twentieth century the waters around Greenland were generally too cold for cod and only small local populations were present in sheltered fjords. However, in 1917 cod from waters off south-western Iceland began to colonise the coast of western Greenland in response to a sudden rise in sea temperatures, amounting to nearly 2°C (Beverton & Lee 1965). The fish spread progressively north as the temperature of the sea rose (Fig. 7.1), reaching Upernavik (72°50′N) in about 1933, although after 1950 the northern limit retreated to Disko Bay (69°N), and this northward expansion was at least partly the result of larval transport due to the increased flow of the Irminger Current (Ford 1978b). This current is an extension of the North Atlantic Drift (see Fig. 7.1) which washes the south-western coast of

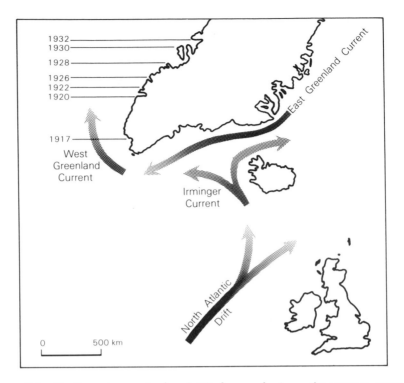

Figure 7.1 Northward spread of cod (*Gadus morhua*) up the western coast of Greenland associated with the warming of the sea and the increased strength of the Irminger Current.

Iceland, to the west of which it splits into westerly and easterly flows. The westerly branch flows back towards Greenland where it joins the cold East Greenland Current to form the West Greenland Current. The intensification of the atmospheric circulation early on in the present century increased the strength of the North Atlantic Drift and hence the vigour of the warm, saline Irminger Current. The greater volume transport of the Irminger Current (Beverton & Lee 1965, Dunbar 1976), coupled with the fact that the Icelandic fish spawned further offshore to the north and west, putting them into the westerly branch of the Irminger Current, resulted in cod eggs and larvae being carried across the Denmark Strait and then round to the waters off the coast of Greenland which were, by then, of sufficiently high temperature to allow the survival and development of the fish. The build-up of the permanent Greenland population was due to exceptionally good survival in a relatively small number of particularly warm years rather than to a gradual change. The year classes appeared in three successive groups: 1917; 1922, 1924, 1926; and 1934, 1936. During the 1930s and early 1940s some mature fish migrated from Iceland to Greenland, but not since 1945 (Cushing & Dixon 1976). There has been little recruitment to the western Greenland stock since 1968 when the cold East Greenland Current became dominant. However, as catches declined in the Iceland–Greenland sector, recruitment increased in the North Sea, where the cod is at its southerly limit, as water temperatures dropped.

It is interesting that the recent decline in the cod stocks off the western coast of Greenland has coincided with the development of a commercial fishery for the Atlantic salmon (*Salmo salar*) in these waters. This fish hatches from eggs and develops in rivers on both sides of the Atlantic (notably in Canada and the USA and Ireland, Scotland, England, France and Spain) and then spends a variable period of time in the sea before returning to fresh water to spawn. Little is known of the marine life of the salmon, although it is possible that they have always fed off the western coast of Greenland. Dunbar (1976), however, suggests that their occurrence there is a consequence of the recent climatic cooling and suggests (Dunbar & Thomson 1979) that salmon were present in these waters during two earlier cool periods: around 1600 and again about 1810.

7.2 Wind transport and terrestrial plants

The spores of bacteria, fungi, bryophytes (mosses and liverworts) and pteridophytes (ferns) are widely dispersed by air currents and, in addition, the fruits and seeds of many angiosperms (flowering plants) are also modified for wind dispersal. Among the latter one might mention plumed seeds (e.g. willow herb (*Epilobium* spp.)) and fruits (e.g. dandelion (*Taraxacum* spp.) and avens (*Geum* spp.)); winged seeds and fruits (e.g. maple (*Acer* spp.)) and woolly seeds and fruits (e.g. willow (*Salix* spp.) and poplar (*Populus* spp.)). Such disseminules will frequently be transported beyond the climatically determined limits for germination and so will die. In climatically favourable individual years, as for example the warm summer of 1976, or during periods of climatic amelioration, such as the early part of the present century, fruits and seeds may be able to develop beyond the species' usual limits and this will result in an extension of the plant's range. Such situations are analogous to that described for the cod above in that both depend upon climatically associated transport in order to take advantage of a climatic amelioration. The corollary is that, without the assistance of such transport, a species may be unable to colonise an area which, owing to a change in climate, has become a suitable environment for it. This accounts for the slow response, or even lack of response, of many species to climatic changes.

Some angiosperms produce what are called 'dust seeds' which, like spores, are minute and extremely light and so tend to be widely dispersed by winds. Such plants (e.g. members of the Orchidaceae, Pyrolaceae and Orobranchaceae) are thus likely to exploit favourable habitats rapidly and so respond quickly to climatic changes. This has been shown by Perring (1965) who, using data on the distribution of plants in the British Isles before and after 1930, presented in Perring and Walters (1962), prepared a list of species which have changed their distributional limits by more than 50 miles (80 km). Leaving out bog and marsh species which have been greatly affected by drainage, of the sixty-three species remaining he found that twenty-nine are dispersed either by spores or light seeds, or are littoral species where dispersal by ocean currents, as described above, is possible. He also showed (see Table 7.1) that the average retreat of these species was greater than for species with no special dispersal mechanisms.

Table 7.1   Movement of distributional limits (pre- and post-1930) of plants in the British Isles (from Perring 1965).

Type of dispersal	Number of species	Average movement (miles)
unassisted	34	86
light seeds	12	94
spores	5	106
littoral	12	147

One example where wind transport of dust seeds appears to have enabled a species to respond rapidly to changes in climate is that of the lizard orchid (*Himantoglossum hircinum*). Although abundant in western France, the only permanent locality where this plant was found in Britain before 1900 was at Dartford, Kent. It had, however, been observed, for short periods only, at about twenty other sites, mainly in the North Downs. These sightings presumably represented plants that germinated in climatically favourable years but were unable to persist to flower in many cases. Between 1900 and 1933, however, a further 129 localities were discovered (see Fig. 7.2), mainly on chalk, although these generally persisted for only a few years. There was a marked decline after 1940 with only twelve localities recorded in the 1950s and only nine since 1960 (Perring 1974). This expansion and contraction of range appears to parallel what is known of the climatic trends in Britain over the same period of time, but without more detailed knowledge of the ecology of *H. hircinum* it is difficult to determine which climatic factors the plant is responding to. The expansion of the species in the early decades of the century has been attributed to the amelioration of winter and spring temperatures and perhaps an increase in the proportion of the annual rainfall falling in winter (Good 1936). However, it is difficult to reconcile this with the fact that since 1940 the plant has persisted in East Anglia where the climatic regime is particularly continental with a tendency to lower winter temperatures and less rainfall than elsewhere in Britain (Ford & Lamb 1976). The greater broomrape (*Orobranche rapum-genistae*) (Orobranchaceae) is another species which produces large numbers of extremely light seeds which are transported by the wind. Before 1930 this species was recorded from 319 ten-kilometre squares but was only recorded from forty-five squares between 1930 and 1950, and from only twenty-two squares after 1950 (Perring 1965, 1974). However,

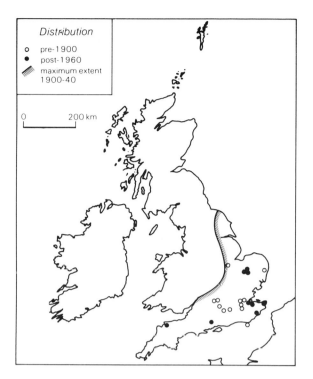

Figure 7.2 Changes in the distribution of the lizard orchid (*Himantoglossum hircinum*). Redrawn from Perring (1974).

although this species has become much rarer, its distribution has not moved in any particular direction which would give a clue as to a climatic effect. The decline of this plant is probably at least partly due to the loss of heathland habitat as it is parasitic on the roots of shrubby Papilionaceae such as gorse or furze (*Ulex*) and broom (*Sarothamnus*).

7.3 Wind transport and animals: insects

It is clear that weather affects a number of aspects of insect flight (see, for example, Johnson 1969) but the question of the relative importance of wind-assisted transport to insect movements is not yet resolved. It is particularly difficult to establish whether wind transport is the main factor accounting for the large numbers of immigrant insects (especially members of the Lepidoptera) in climatically favourable years such as 1945, 1947 and 1976. Amongst the Lepidoptera, the current view appears to be that nocturnal migration of moths is predominantly downwind whilst day-flying butterflies do not appear to orientate their flight with respect to wind direction, but presumably use visual cues which are not available to nocturnal fliers. Taylor,

French and Macaulay (1973) examined records of the silver-y moth (*Plusia gamma*) which after emergence initially flies during both day and night but which confines its flight to night-time after a few days. They found that nocturnal flight is predominantly downwind but diurnal flight is independent of wind direction. These authors consider that direction of flight has little ecological significance in migration as the prime requisite is a general dispersal, in whatever direction. Williams (1958) discounts the influence of wind direction on the movements of both moths and butterflies as they seldom fly more than 10–20 feet (3–7 m) above the ground. He analysed 367 records of butterfly migrations in which both wind direction and flight direction had been reported and found no correlation between the two (Williams 1930). In fact his records can be tabulated as in Table 7.2. Williams draws similar conclusions (Williams 1958) from an examination of 470 flight records of the painted lady butterfly (*Vanessa cardui*) throughout the world and forty-seven flights of the silver-y moth in the British Isles.

Table 7.2   Flight directions reported in 367 records of
butterfly migrations (data from Williams
1930).

Flight direction	Percentage
flying with the wind	22.9
flying diagonally with the wind (i.e. ± 45°)	13.4
flying at right angles to the wind	18.8
flying diagonally against the wind (i.e. ± 135°)	21.3
flying against the wind	23.7

The spectacular butterfly year of 1872 when record numbers of Camberwell beauties (*Nymphalis antiopa*) reached Britain together with our second largest totals of Bath whites (*Pontia daplidice*) and Queen of Spain fritillaries (*Argynnis lathonia*) also provides evidence against the importance of wind transport in butterfly immigration, as these species originate from different directions: Camberwell beauties coming from Scandinavia whilst bath whites and Queen of Spain fritillaries come from central and southern Europe.

In the past it has been suggested that Camberwell beauties are brought to England in cargoes of timber from the Baltic, although this is no longer considered likely. Ship-borne transport may, however, be a factor in the periodic arrival of monarch or milkweed butterflies (*Danaus plexippus*) in Britain. This American species is occasionally found along the southern coasts of England and Wales in autumn when, in its home country, it is migrating south to hibernate in Florida and Mexico. It may be accidentally carried out to sea and then transported across the Atlantic on the prevailing westerlies; however, if this were the case one would expect more records from the western coast of Ireland, though there are in fact only three. This may of course reflect lack of observers rather than lack of butterflies. The concentration of sightings of this species in the south of England has led to the suggestion that the butterflies are in fact carried across the ocean on trans-Atlantic ships and then fly a comparatively short distance to the point at which they are sighted. However, in Britain there are occasional records from the northern part of England and one record from the Shetlands. Furthermore, if ship transport were implicated one would expect similar sightings from continental Europe (e.g. from around Cherbourg) but the butterfly has only been recorded there six times: twice in France and four times in the Iberian Peninsula. This may not be a valid criticism as ships bound for France would still pass close to south-western England so that if the butterflies flew towards the first accessible piece of land, possibly assisted by on-shore breezes, most of them would end up in Britain. At present it is not possible to reach a conclusion on this matter.

There is some evidence that the American moth *Phytometra biloba* can be carried across the Atlantic by winds. Specimens of this species are only rarely found in the British Isles and one such occurrence was noted at Aberystwyth, Wales, on 19 July 1954. Examination of the weather maps for the preceding period suggested (Hurst 1969) that a flight over the Atlantic was possible in warm air south of the polar front. Active depressions were moving across the Atlantic and the flight would have taken $3\frac{1}{2}$ days, although, once again, the possibility of ship-borne transport can not be excluded. There are a few examples of unusual immigration of other species of moth to this country which can be fairly conclusively related to particular synoptic weather conditions. One case is that of the enormous immigration of the diamond-back moth (*Plutella maculipennis*) along eastern and north-eastern coasts of Britain in June 1958 (French & White 1960, Shaw 1962). This species is indigenous but only rarely builds up sufficient numbers to develop as a serious pest on cruciferous crops; such build-up generally being associated with a sudden immigration. As this moth is extremely small (with a wing span of about 16 mm) it was thought likely that the 1958 immigration was wind-assisted. Analysis of the appropriate weather maps indicates that the moths probably originated from the eastern shores of the Baltic: there was a generally easterly airstream approaching the British Isles between a high to the north and north-east and a low to the west and south-west. Other notable immigrations of this species occurred in 1891 and 1914. The predominance of westerly conditions may have accounted for the lack of such phenomena between 1914 and 1958.

Another small (about 20 mm across) moth, the small mottled willow moth (*Laphygma exigua*), whose larva is an agricultural pest in northern and central Africa, is found in small numbers each year in Britain. 1962, however, was an outstanding year with nearly 1200 specimens being captured. The main immigration took place on 6 May and back-tracking along the prevailing south-south-westerly airstream (Hurst 1963) indicated that the moths had left Morocco about 4 days earlier (see Fig. 7.3). Larger than normal catches (over 100) of this moth in 1952 and 1958 (Hurst 1965) can also be explained by wind-assisted transport from Morocco and Madeira respectively.

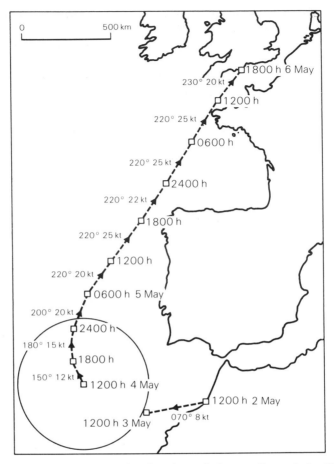

Figure 7.3 Back-track of the immigration of the small mottled willow moth (*Laphygma exigua*) on 6 May 1962. At each point the time and day of the month are given on the right and the wind direction and speed in knots on the left. Beyond a point west of Gibraltar the winds were light and variable in an anticyclonic area marked by the circle with radius 200 miles. Redrawn from Hurst (1963).

Before leaving the Lepidoptera, it should be mentioned that some species have become structurally modified so as to reduce the possibility of wind transport: the development of dwarf races less susceptible to the influence of wind enables species to exist in habitats with markedly oceanic regimes. For example, the subspecies *caernensis* of the silver-studded blue butterfly (*Plebejus argus*) is a dwarf race (the wingspan is about 4 mm less than the typical form) associated with coastal limestone in Caernarvonshire, Wales. The dwarf race *thyone* of the grayling (*Hipparchia semele*) is also found in North Wales, on Great Orme's Head. In both species the dwarf form emerges several weeks earlier than the typical one as its smaller ultimate size reduces development time.

Of the forty-three species of British dragonfly, about ten are known to be migrants, the most common being *Libellula quadrimaculata* and *L. depressa*. These migrants all belong to the sub-order Anisoptera; none of the sixteen British members of the sub-order Zygoptera, or damselflies, is known to migrate. A possible example of the effect of prevailing wind patterns on dragonfly migration is the arrival of large numbers of *Sympetrum striolatum* along the southern coast of Ireland, starting on 2 September 1947. As no immigration was reported along the French coasts (or even from Devon and Cornwall) it has been suggested (Longfield 1948) that the dragonflies had travelled more than 500 miles (800 km) over the sea from Spain and Portugal on the prevailing winds.

The locust swarms found in arid sub-tropical regions generally move downwind (although, within a swarm, individuals may move in any direction with respect to the wind) as this brings them to a convergence zone, such as the Inter-Tropical Convergence Zone (see Sec. 3.3), characterised by high rainfall and hence newly-flushed vegetation (Rainey 1951, 1973). The desert locust (*Schistocerca gregaria*) has only been reported from Britain twice (Williams 1958). In October 1869 a number were found over south-eastern England and these represented the first records from anywhere in Europe. On 17 October 1954, two more specimens were captured in the Scilly Isles and one offshore at Tramore in Co. Waterford, Ireland. Six further specimens were subsequently reported from the Scillies, southern Ireland and Cornwall. In October 1954, at least, and possibly in 1869 as well, a strong southerly airflow in the eastern Atlantic presumably carried these insects out to sea off the north-western coast of Africa from their origin in Morocco, and then north towards southern England and Ireland.

Reports of invasions by migratory locusts (*Locusta migratoria*) in various parts of Europe, and also Britain, can be traced back to at least the ninth century (Müller 1953, Rudy 1925) and provide information concerning abnormal incidence of southerly and easterly winds in summer, useful in reconstructing climatic history (see, for example, Lamb 1977a, p. 234). These reports presumably refer to insects which had originated from the permanent breeding ground of this species on the north-western shores of the Black Sea. In recent times Williams (1958) reports 10 years between 1900 and 1940 when individual migratory locusts were reported from Britain, occasionally as far north as Orkney and the Shetlands. In 1931 there were many records over eastern England as far north as Yorkshire. No breeding was reported anywhere in Europe, apart from the Black Sea, until 1944, when large numbers were discovered in the Gironde district of south-western France. A serious outbreak developed in this region during 1945 and 1946 when about ten locusts were recorded from the southern coast of England. In 1947, thirty-two were caught in England, mainly along the southern coast again, but one specimen was reported from Lincolnshire.

It is clear that downwind migration predominates in small insects such as aphids and thrips which do not have the physical strength to fly against the wind. Although aphids are believed to fly for only short periods at a time, the significance of wind-assisted transport in these insects was first established in 1924 when Charles Elton captured spruce aphids on the treeless islands of Spitzbergen (Elton 1925). These aphids had been carried across the Arctic Ocean from the Kola Peninsula in north-western Russia, a distance of more than 800 miles (1300 km). Sudden increases in the number of aphids caught in traps in Britain, such as those of the Rothamsted Insect Survey, may represent influxes from the continent. This is certainly a possibility with cereal aphids where the crops further south in Europe would be more advanced than those in England, and examination of weather maps has indicated that at least some cases of high numbers of the English grain aphid (*Sitobion avenae*) are associated with circulation features consistent with a continental origin. Similarly, Hurst (1969) concluded from an examination of synoptic charts that exceptionally large numbers of the peach-potato aphid (*Myzus persicae*) captured in July 1947 were probably the result of wind-assisted immigration from Germany or Belgium. There is, however, little direct evidence of transport of aphids from Europe, although there is a report (Hardy & Milne 1937) of considerable numbers of live aphids captured over the North Sea, some 120–150 miles (190–240 km) from the nearest land.

   Similar uncertainties surround the origins of the large numbers of aphid predators, notably hoverflies (Syrphids) and ladybirds (Coccinellids), that are found along British coasts from time to time. In 1975 the syrphid *Syrphus corollae* was very abundant, whilst in 1976 the 7-spot ladybird *Coccinella 7-punctata* was the predominant species. In 1977 syrphids were again very abundant, although on this occasion the main species was *Syrphus balteatus*. As larvae, all these species feed on aphids and their abundance in 1975–77 reflected the high numbers of aphids in these years. However, in each year, large numbers of the relevant predator have been found along the tide-line of beaches along the east and southern coasts, suggesting that immigration from the continent may have been partly responsible for their rapid response to the aphid population increase. Thus in a letter to *The Times* newspaper of 25 August 1977 Dr C. D. Putnam wrote: 'At Southwold on August 9 large numbers of three species of hoverfly were quite definitely flying in, low over the sea, from 9 am to 3 pm, flying against a fresh offshore breeze'. Southwold is on the Suffolk coast at 52°20′N 1°40′E. Dr Putnam goes on to say that he counted over 1000 dead flies per foot of tide-line over a distance of 3 miles, equivalent to fifteen million dead flies, these presumably being the weaker insects that were unable to complete the crossing successfully. In a letter to *The Times* of 29 August 1977 Dr W. S. Bristowe recorded 'vast swarms' of *Syrphus balteatus* and *S. ribesii* between 8 August and 18 August from Devon to East Anglia and suggested that they had been carried across from Holland and France. The contrary view is that these are native-born insects that are carried out to sea and then washed or blown back again, although this does not satisfactorily explain Dr Putnam's report above. It is difficult to resolve the question of the origin of these insects without reports from ships.

The large number of 7-spot ladybirds (*Coccinella 7-punctata*) in 1976 was widely observed throughout the country (Owen 1976) and was clearly not primarily due to immigration but was the result of successful overwintering with subsequent rapid reproduction in response to an abundant supply of aphids as food. However, this species was also found in large numbers along coastal beaches and, together with other members of the Coccinellidae, is certainly known to migrate within Britain, so once again immigration from the European continent is a possibility. However, another suggested explanation for their being found in large numbers at the coast is that as they have a high moisture requirement, in times of drought they accumulate on beaches to drink the sweat on the exposed bodies of sunbathers. In fact in 1976 there were many reports of sunbathers being 'bitten' by ladybirds which were probably due to disturbed insects employing their usual defence mechanism of exuding caustic fluid from their leg joints.

One beetle that certainly appeared to have crossed from the European continent aided by wind transport in 1976 was the Colorado beetle (*Leptinotarsa decemlineata*), which can devastate potato crops. A breeding colony was discovered on a farm near Thanet, Kent, in September 1976, the first outbreak in Britain since 1952, and it was thought that the beetle had been blown over from the continent where it is prevalent. Past occurrences of this beetle have been associated with wind-borne transport as the insect is unable to fly long distances unassisted (Hurst 1970).

7.4 Wind transport and animals: birds

Owing to their greater size and strength, the flight direction of birds is less susceptible to the influence of winds than is that of insects. Even small passerines are able to fly against the wind during migration, although this appears to be a temperature response: the migrants orientate to the warm air carried by the wind rather than to the moving air current itself (Dorst 1962). Other species make use of the winds flowing around major circulation features, such as depressions or highs, when migrating. For example, Schenk (1924) showed that the arrival of the woodcock (*Scolopax rusticola*) in Hungary in spring was always associated with warm south-easterly winds blowing anticlockwise around a depression situated to the north-west, usually centred over the British Isles.

This anticlockwise cyclonic approach is characteristic of the autumn migration of a number of birds from Greenland and Iceland and often involves a very long, although wind-assisted, non-stop flight over the sea which can exhaust the food reserves of the bird undertaking it. For this reason, the Greenland races exploiting this technique are frequently larger than the continental European forms of the same species (Williamson 1958). For example, the Greenland subspecies (spp. *leucorrhoa*) of the wheatear (*Oenanthe oenanthe*) is much heavier than its British or continental European counterparts as its cyclonic approach to Britain, *en route* to West Africa, takes it around the western and southern edges of the large depressions moving across the Atlantic. This involves a non-stop flight, albeit assisted by a 50 knot wind, of between 1500 and 2500 km entailing a loss of body weight of up to one-third (Williamson 1976). Occasionally birds undertaking a cyclonic approach may be carried further than usual and die as their food reserves become exhausted. This happened in autumn 1959 with the Iceland race of the redwing (*Turdus iliacus coburni*), which usually winters in western Scotland

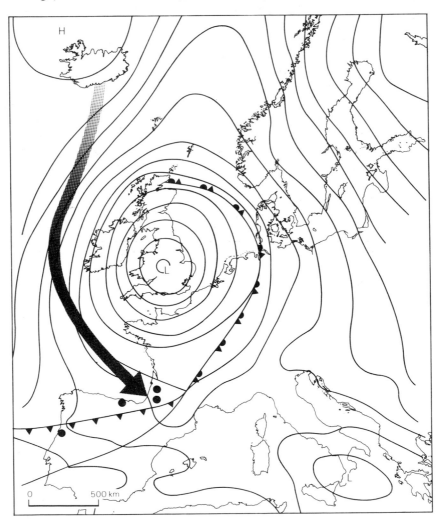

Figure 7.4 Recoveries of ringed specimens of the Icelandic race of the redwing (*Turdus iliacus coburni*) in France and Spain (recoveries indicated by black dots) following cyclonic approach around a deep depression on 14/15 November 1959. Redrawn from Williamson (1976).

and Ireland. Birds of this race which had been ringed in Reykjavik were found dead or dying from exhaustion in southern France and northern Spain between 15 and 17 November (Williamson 1969). Just prior to this a very large depression was centred over the British Isles and the birds were directed around this by the prevailing winds and thus travelled about twice as far as usual (see Fig. 7.4).

Some birds may be carried right across the Atlantic in depressions and this probably accounts for the origin of the north-eastern Atlantic colonies of Leach's storm petrel (*Oceanodroma leucorrhoa*), which is predominantly an American bird. This species winters south-east of the Caribbean and occasionally depressions derived from hurricanes in this area spill hundreds of trapped birds over the British Isles as in the 'wreck' of Leach's petrels in October 1952 (Boyd 1954). There is, in fact, a continuous transport of birds across the Atlantic Ocean on the prevailing westerlies (although, as with butterflies, Durand (1972) has pointed out that ship-assisted transport is a possibility) and the latitude of arrival of these American vagrants will vary with the expansion and contraction of the circumpolar vortex. Most of the arrivals occur in autumn when birds in eastern North America are migrating south to their overwintering areas. Table 7.3, adapted from Williamson

Table 7.3 Proportion of nearctic species among new birds species admitted to the British and Irish lists (data from Williamson 1974).

Decade	Number of species	Proportion nearctic %
1870–79	4	50
1880–89	7	14
1890–99	8	13
1900–09	18	28
1910–19	6	0
1920–29	4	50
1930–39	4	25
1940–49	4	0
1950–59	27	59
1960–69	23	44

(1974), shows how the proportion of American or nearctic species among the new birds admitted to the British and Irish lists since 1870 has varied in each decade.

It is perhaps dangerous to draw conclusions from the variations in the number of new species decade by decade as bird-watching has become especially popular in recent years while, on the other hand, there was clearly a decline in ornithological activity associated with the two World Wars. However, the dominance of nearctic species since about 1950 is marked and numbers of individuals have been high as well (Elkins 1979) with over 130 individuals of thirty-two American species during the three years 1960–62 (Williamson 1974). The pectoral sandpiper (*Calidris melanotus*) is the most frequent vagrant with an average of twenty specimens a year. Another example is Wilson's phalarope (*Phalaropus tricolor*), which was admitted to the British list in 1954 but was only irregular until 1960. It has occurred annually since with a total of sixty-seven birds by 1975 (Williamson 1976) and a maximum of ten individuals in 1971 (Smith 1972).

This recent increase in the number of American vagrants over the past two decades can be associated with a southerly movement of the mean path of the Atlantic storm-track. During the climatic amelioration of the early part of this century, the prevailing centre of low pressure was near to or north of Iceland. In consequence, the mean path of the North Atlantic depressions ran from south-west to north-east so that most trans-Atlantic vagrants would reach Greenland (Williamson 1974). With the reduction of the vigour of the atmospheric circulation in recent decades, the centre of low pressure has moved 2–5° of latitude further south (Lamb 1963, Perry 1971) so that most vagrants now arrive in southern Ireland, Brittany or south-western England. Similarly, the great skua (*Stercorarius skua*), whose oceanic distribution is determined by the location of the prevailing westerlies driving the North Atlantic Drift, has increased in northern Scotland in recent years as birds have been displaced south from Iceland.

There are two well documented instances of wind-assisted trans-Atlantic transport in the opposite direction, i.e. from Europe to America, both involving the lapwing (*Vanellus vanellus*) (Witherby 1928, Williamson 1969). In each case the mechanism involved a cyclonic approach as there was an extremely large depression situated far south in the Atlantic in mid-winter, producing a continuous easterly airstream right across the Atlantic on its northern edge. A similar situation in the eastern Atlantic in winter 1936 displaced fieldfares (*Turdus pilaris*) on their autumn migration from Scandinavia across the North Sea to Britain so that they were carried north and then west to Greenland, where they founded a breeding colony (Salomonsen 1951).

In general, birds migrating from Scandinavia to Britain in the autumn do not fly directly across the North Sea. They usually wait for a spell of anticyclonic weather when visibility is clear and winds are light, and then skirt the North Sea from Denmark to northern France where they drift across to England on the easterly winds on the southern edge of the high pressure system. In autumn 1951 there was an example of such anticyclonic drift on a large scale when an anticyclone intensified over Scandinavia at the end of September and large numbers of small birds, notably robins (*Erithacus rubecula*), poured into eastern England over several days (Williamson 1952). As with cyclonic transport across the Atlantic, late summer anticyclonic development over Siberia can result in Far Eastern migrants being carried westwards instead of travelling south to their winter homes (Williamson 1969). In addition, in recent decades the predominance of north-easterly winds associated with the increased frequency of blocking anticyclones in winter has resulted in occasional influxes of northern species to lower latitudes. A notable example was the arrival of large numbers of nutcrackers (*Nucifraga caryocatactes*) in north-eastern Britain in 1968, carried there on a north-easterly airstream from their usual homes in Russia and southern Scandinavia.

High latitude anticyclones in spring may deflect returning migrants westwards and this effect appears to be implicated in the recolonisation of Scotland by the osprey (*Pandion haliaetus*). This fish-eating bird nested at a number of sites in Scotland until the end of the nineteenth century when it declined rapidly due to over-hunting for the millinery trade. The last successful breeding anywhere in Scotland was in 1916. In 1954, however, a pair nested at Loch Garten, Inverness-shire, and the number of birds and nesting-sites have increased over successive years with a record twenty-two pairs nesting in Scotland in 1979 – the highest number recorded this century. This recolonisation has been associated with a tendency for a blocking anticyclone to become established over Greenland or Scandinavia in the spring. As a result, birds migrating in April from their wintering sites in northern Africa back to their original Scandinavian homes have been deflected to the west by the easterly winds flowing around the southern side of the anticyclone and have ended up in Scotland. The population in Scotland must be maintained by a continuous annual input of Scandinavian birds as it is unlikely that any young survive to return to Scotland. There is evidence from ringing studies (Mead 1973, Williamson 1976) that a high proportion of young birds are shot as they fly over France and Spain. Thus the intensive measures taken at Loch Garten to protect the ospreys are rendered ineffective once the birds leave the area.

Westward anticyclonic drift during spring migration probably accounts for the recent nesting of the wryneck (*Jynx torquila*) in Inverness-shire and Ross-shire (Burton, Lloyd-Evans & Weir 1970) and of the red-backed shrike (*Lanius collurio*) in Orkney (Balfour 1972). Both these species have been declining in southern England and their presence in Scotland, where they have never been known to breed before, indicates that these are birds of European origin.

Another feature associated with anticyclones but not an effect of wind transport is the phenomenon of 'overshooting' on spring migration. In this situation, birds are stimulated by the favourable conditions prevailing during anticyclonic weather to continue their migration further north than usual and overfly their usual summer homes. This probably accounts for the colonisation of Iceland by birds such as the black-headed gull (*Larus ridibundus*), the swallow (*Hirundo rustica*) and the starling (*Sturnus vulgaris*) during the twentieth century climatic amelioration, as well as the colonisation of Shetland in 1953 by the rook (*Corvus frugilegus*), which built up a considerable population in subsequent years at a single suitable locality.

8
Climatic effects mediated by pathogens

8.1   Introduction

Knowledge about the biology of the pathogens (causative agents of disease) of wild animals and plants is very sparse; most available information relates either to humans or agricultural animals and crops. However, the fact that certain diseases are prevalent at particular times of the year suggests an underlying climatic influence. In many cases, however, the mechanism of such an influence is difficult to establish. For example, is the higher incidence of respiratory infections in man in winter due to a positive effect of cold weather on the causative pathogens or to the depression of the immune response at lower temperatures; or perhaps it is related to the increased opportunities for transmission as people are confined indoors for longer periods, or even to the lack of vitamin C due to the reduced availability of fresh vegetables? More information about the influence of climate on diseases, part of the subject of biometeorology, may be found in Tromp (1963) and Sargent and Tromp (1964).

Climate can indirectly influence the incidence of pathogen attack by reducing host resistance. In such situations climate is acting in the same fashion as any other stress-producing factor which weakens an organism and so permits the invasion of pathogens. Mattson and Addy (1975) quote a number of examples where outbreaks of forest pests occurred after trees had been subject to stress such as excessive moisture, drought or pollution, and a similar phenomenon was found with diseases of trees after the 1976 drought in Britain. Dutch elm disease (see later) killed more trees than usual as they were weakened by water-stress and beech bark disease was also prevalent and particularly affected those trees on freely draining chalk soils. This drought year also provided an interesting demonstration that there are sometimes several links between a climatic influence and the response of a pathogen when the hot summer resulted in the development of large numbers of the English grain aphid (*Sitobion avenae*) on cereals. The feeding activities of these aphids resulted in the production of large quantities of sugar-rich honeydew which then encouraged the development of sooty mould fungus on the crops.

8.2 Climate and the spread of disease: activity of vectors

Many pathogens are transmitted by animal vectors, usually insects, and the effect of climate upon the activity of these organisms (such as described in Sec. 6.7) can influence the rate of infection and the size of the infected area. An important example from human history is the case of bubonic plague or the Black Death which is caused by a bacterium (*Pasteurella pestis*) transmitted by fleas of the black rat (*Rattus rattus*). The rat flea (*Xenopsylla cheops*) is particularly active at temperatures between 20 and 32°C, which helps to explain the association between plague outbreaks in England in the seventeenth century and hot summers. One of the most notable outbreaks during this period was that in London during the hot summers of 1665 and 1666 which was brought to an end by the Great Fire (the rapid spread of the fire itself was presumably related to the dry condition of the timber of houses resulting from the heat and drought of 1666). Ague or malaria is another human disease which is transmitted by an insect vector, this time the mosquito *Anopheles*. Apart from the necessity of stagnant water for egg-laying, the mosquito requires high temperatures for breeding and hence the 16°C July isotherm marks the northern limit of malarial districts. The parasite itself, the protozoan *Plasmodium vivax*, generally requires somewhat higher temperatures of the order of 19–20°C in order to develop within the lifetime of the mosquito (McNalty 1943). These requirements for high temperatures perhaps explain why ague or malaria reached its peak in England in the mediaeval warm period.

Dutch elm disease has recently become a severe threat again to elm trees in Britain and Europe due to the importation from Canada in the late 1960s of a new 'aggressive' strain of the causative fungus (*Ceratocystis ulmi*). The disease was originally detected in Europe in 1918 and had probably come initially from Asia. It was first reported in south-eastern Britain in 1927 and caused widespread death of elm trees until it declined after about 1937, although remaining present at low levels. The disease was carried to North America in imported logs in the 1930s and caused a major environmental change as it spread across the continent, eventually developing into the new aggressive strain which was again transported across the Atlantic in shipments of logs. The current British outbreak has caused the death of millions of trees (Clouston & Stansfield 1979) and has moved progressively northwards from what must have been its ports of entry. The disease has had a major impact on the more than forty species of invertebrates entirely restricted to elm, such as the white-letter hairstreak butterfly (*Strymonidia w-album*), as well as on birds such as the rook (*Corvus frugilegus*) and owls (barn owl (*Tyto alba*) and tawny owl (*Strix aluco*)) which used to nest in elm trees.

The fungus induces the growth of balloon-like protrusions called tyloses across the water-carrying xylem vessel elements and so the tree dies from lack of water, an effect which is clearly enhanced in drought years such as 1976. Although the disease can be transmitted along hedgerows via the roots of infected trees, the principal mechanism of dispersal is by means of attachment of spores of *C. ulmi* to bark beetles of the species *Scolytus scolytus* and *S. multistriatus*. Adults of these species lay eggs in breeding galleries under the bark of elm trees and the larvae which hatch out make secondary galleries at right angles to the primary gallery, producing a characteristic pattern of tunnels. The larvae pupate and emerge between May and October. The beetles will only fly in dry weather when the temperature is above 15–16°C and they are more active in warm summers. It has been shown experimentally (Gibbs 1974) that infection of a tree in May or June is particularly effective as in these months the fungal spores can enter the springwood vessels and spread rapidly, so warm, dry conditions in these months would be particularly deleterious. The increased activity in the extremely warm summers of 1975 and 1976 was of considerable importance in accelerating the spread of the disease. An analogous case is that of the house longhorn beetle (*Hylotrupes bajulus*) which is a pest of the sapwood of conifers such as those used in roofing supports and is unusual in being able to breed in fully seasoned wood. Adult flight requires temperatures above 16.5°C and infestations are confined to periods with high summer temperatures. The species caused considerable damage to woodwork in houses in late Georgian times and then died out until becoming a serious pest again during the warm summers of 1934–53 (White 1954).

Although many pathogen vectors are insects, other types of animals may be involved in disease transmission as in the case of fascioliasis which has as a vector the snail *Lymnaea truncatula*. Fascioliasis is the disease resulting from the infection of sheep and cattle with the liver fluke (*Fasciola hepatica*). Eggs pass out of a parasitised animal in the faeces and then hatch to produce a ciliated miracidium larva which must bore through the integument of a *L. truncatula* snail within 24 h. The miracidium requires temperatures above 10°C and a waterlogged environment in which to swim and these conditions are also optimal for the breeding of the snail vector. Fascioliasis is endemic in the west of England and Wales where rainfall is higher, and the disease is especially prevalent during wet summers.

Some diseases, especially those transmitted by viruses, may be dispersed by wind rather than the activities of vector organisms. This is the case with Fowl pest which is a respiratory disease of birds caused by a virus. It has been shown (Smith 1964) that wind-borne dispersal of this virus could account for some 85% of outbreaks over distances up to 50 km from a source of infection. The possibility of foot-and-mouth virus being spread by the wind was first suggested by Hardy and Milne (1938) who proposed that it was carried on particles and insects, and by McClean (1938) who suggested that it might be absorbed on aeroplanktonic yeasts. In fact, diseased animals exhale the virus in droplet nuclei which evaporate leaving the virus in a particle which will travel freely. The 1967–68 outbreak in Britain (Hugh-Jones & Wright 1970) started in the West Midlands in autumn 1967 and the initial stages of spread were marked by wet conditions with continuous south-westerly winds which spread the disease downwind (Smith & Hugh-Jones 1969). There is evidence of transmission of the virus by wind over distances up to 150 km and Hurst (1968) has shown that many of the primary outbreaks which have occurred near the eastern and southern coasts of England can be attributed to wind-borne spread from the Continent as happened in March 1981 when an outbreak occurred on the Isle of Wight.

8.3 Fungus diseases of plants and animals

Many fungus spores will only germinate when immersed in water, although for those of the powdery mildews humid air is all that is required. Thus, outbreaks of fungal diseases of plants are usually associated with wet or humid conditions providing a layer of moisture on the host leaf surface through which the germ tubes of settling spores can penetrate. A notable example is the potato blight fungus (*Phytophthora infestans*) which requires periods of over 48 h during which air temperature remains continuously above 10°C with relative humidity above 90% for successful development (Smith 1956, Smith & Walker 1966). In Ireland, the warm, wet summers of 1845–48 (especially 1846 and 1847) provided ideal conditions for the spread of the fungus which had recently been introduced from America in infected tubers.

Many rusts and mildews are encouraged by wet summers, for example wheat stem rust requires surface moisture together with temperatures of 18–24°C for successful infection of young plants. Ergot blight (*Claviceps purpurea*) on rye can produce haemorrhages and hallucinogenic symptoms as well as spontaneous abortion in humans who eat bread made from infected grain. The blight develops in damp weather and the various outbreaks of human ergotism or 'St Anthony's Fire' ('Fire' from the blackened, gangrenous limbs of the sufferers of the disease and 'St Anthony's' after the Hospital Brothers of St Anthony, an order founded in 1096 to care for such people) in the Middle Ages in Europe can be associated with periods of more maritime westerly weather (Bryson & Murray 1977). A modern outbreak in Russia in the 1940s was attributed to the eating of bread made from grain stored through the winter under snow and there, in the damp conditions, becoming infected with blight. In cattle, bovine mycotic abortion is caused by the presence of *Aspergillus* mould on their hay or straw and its incidence is correlated with the moisture content of the mature hay, i.e. with the rainfall in June, the main haymaking month. Farmer's lung or allergic alveolitis, a pulmonary condition resulting from the inhalation of minute particles of mouldy hay or grain containing spores of *Thermopolyspora polyspora,* is similarly prevalent after a wet June, although the spores are liberated in dry weather.

Among trees, cool moist springs (such as that of 1977 in Britain) are ideal for infection by the fungal spores causing anthracnose disease in weeping willow (*Salix babylonica*), London plane (*Platanus* X *hybrida*), walnut (*Juglans regia*) and poplar (*Populus* spp.). Unlike Dutch elm disease, anthracnose disease is not fatal but causes spotting and browning of leaves with some dieback of twigs. In contrast, sooty bark disease of sycamore (*Acer pseudoplatanus*), bark necrosis of beech (*Fagus sylvatica*) (a fungal (*Nectria* sp.) disease that enters the tree through holes made by a phloem-sucking coccid insect, *Cryptococcus fagi*) and oak mildew on *Quercus* are always more serious after hot dry summers and were prevalent in Britain in 1976 due to the hot conditions in that and the previous summer. Oak mildew (*Microphaera alphitoides*) was also widespread after the hot dry summers of 1921, 1955 and 1959 and sooty bark disease (caused by the fungus *Cryptostroma corticale*) was prevalent after the warm summers of 1947 and 1959. Recently, considerable anxiety has been expressed over the deaths of many cypress trees in Italy, especially those so characteristic of Rome. This appears to be due to a bark-invading fungus, *Coryneum cardinale* (an aphid, *Cinara cupressi,* has also been implicated, but it seems likely that the aphids are just more abundant on those trees which are in poor condition due to fungal infection), and there have been a number of suggestions that the outbreak is climatically induced in some way.

8.4   Diseases of insects

Although comparatively little is known about the diseases of insect groups other than in the context of the biological control of pest species, there are a number of examples among the Lepidoptera where the influence of climate upon the incidence of disease appears to limit the distribution of a species. An interesting case is that of the black-veined white butterfly (*Aporia crataegi*), which was locally abundant over a large part of southern Britain in the nineteenth century but became extinct by 1925. A number of unsuccessful attempts were subsequently made to reintroduce it in Kent, its last foothold in England. The larvae of this butterfly hatch in August and are gregarious on blackthorn (*Prunus spinosa*) hedges and orchard trees until they hibernate in October, and it has been suggested that the milder, moister weather associated with the more oceanic conditions prevailing in the early decades of the present century encouraged a virus or fungal disease which rapidly spread through these communal larvae. There appears to be no positive evidence for this suggestion, although it is interesting that the last specimens of *A. crataegi* were taken in eastern Kent, where the oceanic influence of the prevailing westerlies might be expected to have been less pronounced. Furthermore, Ekholm (1975) reports that in Finland, where occasional immigrants breed, the larvae can withstand very cold temperatures but not moist winter weather.

Two species of the genus *Colias* are frequently taken as continental immigrants to Britain: *Colias croceus,* the clouded yellow, and *Colias hyale,* the pale clouded yellow. In warmer countries *C. croceus* breeds continously throughout the year and has no hibernating stage, so would probably be unable to survive the colder winters of Britain. In contrast, the larva of *C. hyale* does hibernate and so should, in theory, be capable of overwintering in

England, but there is little evidence of this occurring. It had been suggested that the larva was unable to withstand temperatures below 40°F (4.4°C) but E.B. Ford (1957) reports that Mr J. Shepherd twice brought the larvae through an English winter without artificial heat, and on one occasion they survived a temperature as low as 19°F (-7.2°C). Thus, it may not be low temperatures that prevent the overwintering of the caterpillar, but perhaps the damper conditions prevailing in the more maritime British Isles may make the species more liable to fungal or other infection. If this were so, mild winters would perhaps be more deleterious than cold winters. A similar phenomenon may explain why the Camberwell beauty butterfly (*Nymphalis antiopa*), although well able to withstand the rigours of a Scandinavian winter, rarely hibernates successfully in Britain.

Amongst other insect groups, the red ant *Formica lugubris* has a much more northerly distribution than the closely related *Formica rufa*. In Britain it is widespread in Scotland and the uplands of Wales and northern England but also occurs in southern Ireland. Satchell (1965) states that between the mid-nineteenth century and the 1920s it was recorded from seventeen sites in Ireland, although by 1963 only the most northerly colony survived; no colony

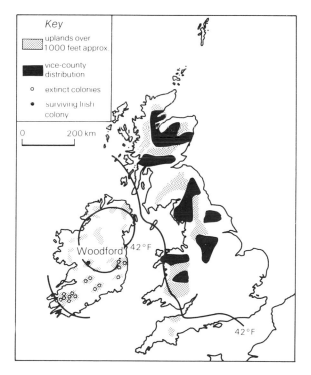

Figure 8.1 Distribution of the ant *Formica lugubris* in 1963, showing the extinction of all the Irish colonies lying on the warmer side of the 42° F mean January isotherm for the period 1901–30. Redrawn from Satchell (1965).

extinctions were reported from mainland Britain. Figure 8.1 shows that all the extinct colonies lie on the warmer side of the 42°F (5.6°C) January isotherm (1901–30 mean) while the surviving Irish colony and the mainland colonies are on the colder side. As the extinction of these colonies coincided with a period of warmer winters, it is possible that the milder conditions favoured the development of some infective pathogen. However, Collingwood (1958) suggested that high winter temperatures impair the production of sexual individuals. He quotes Stelfox as being unable to find any alates (winged sexual forms) in colonies in Wicklow visited fairly regularly in the years 1921–26 and Gösswald and Bier (1954) who demonstrated that colony groups of *F. rufa* did not produce sexuals when kept under constant temperatures whereas *F. nigricans*, which extends to the Mediterranean area, developed normally under the same conditions. It is, however, perhaps unwise to draw definite conclusions from the work of Gösswald and Bier as constant temperature conditions are never experienced in the wild. If climate is implicated in the decline of these colonies, it must have been acting in concert with some other deleterious factor or the species would have become extinct during the mediaeval warm period (see Sec. 4.5).

8.5   Diseases of vertebrates

The influence of climate on the incidence of disease in vertebrates is very imperfectly understood, even in the case of man. This is at least partly attributable to the fact that many vertebrates are warm-blooded and so buffered, to some extent, from the effects of climate. Furthermore, in attempting to unravel the causes of disease it is extremely difficult to interpret the relative significance of prevailing climatic conditions on species which, unlike for example the hibernating stages of an insect or a rooted plant, are relatively mobile and may experience considerable variations of climatic parameters over a short period of time. In aquatic habitats, however, temperature is the only climatic factor of predominant importance and in a given body of water temperature tends to be spatially coherent. Thus there is some evidence that the changing climate has brought about changes in the incidence of diseases in such habitats. A case in point is the recent epidemic of ulcerative dermal necrosis which originated in the 1960s.

Ulcerative dermal necrosis or UDN is a disease of salmonid fish, characterised by grey circular necrotic ulcerations which start on the head of the animal and spread to other unscaled areas of the body. The animal suffers progressive weakening because of diffusion of blood proteins through the wounds into the surrounding water, and passage of water in the opposite direction into the fish gives the flesh a waterlogged quality. Post-mortem examination often shows that the gall-bladder is affected. The aetiology of the disease is uncertain and, although Stuart and Fuller (1968) believe that a *Saprolegnia* fungus is the causative organism, much of the evidence suggests that a virus is the primary cause with secondary infection by fungi such as *Saprolegnia ferax* and bacteria such as *Chondrococcus (Cytophaga) columnaris.*

The current epidemic was first noticed in a number of rivers in south-western Ireland in 1964. During 1966 it spread to Lancashire, Cumberland and the Solway river system and is now widespread in the rivers of England, Wales and Scotland. The previous outbreak started in the Solway in the 1870s and seems to have lasted until about 1905.

The prevalence of the disease is increased during the autumn and early winter and Elson (1968) reports that a comparison of the incidence of the disease in one river system with temperature records suggests that it is most common when the water temperature is between 7 and 10°C. This apparent association with lower temperatures may explain the absence of the disease during the first half of the twentieth century. Further research on the relationship between river temperatures and the incidence of this commercially important disease should be undertaken. It is interesting that a similar skin infection has been reported in flatfish, including the sole (*Solea solea*) in severe winters such as 1946/7 (Simpson 1953) and 1962/3 (Crisp *et al.* 1964) and also a cold spell in March–April 1929 (Lumby & Atkinson 1929).

Another vertebrate disease that occurs in aquatic environments and has a climatic dimension is botulism. This is a paralytic disease resulting from the ingestion of the neurotoxin produced by the anaerobic bacterium *Clostridium botulinum* and has resulted in the deaths of considerable numbers of wildfowl in recent years. There appear to be seven forms of the bacterium, which are termed Types A to G. Type A gives rise to human botulin food poisoning which was originally called botulism because of its frequent association with the consumption of sausage or, in Latin, *botulus* (Smith 1976). Type C is the main cause of bird mortality, although it also affects cattle and other mammals. The third major form, Type E, is thought to have been implicated in some outbreaks of avian botulism on Lake Michigan (Fay 1966).

The spores of the bacterium occur naturally in soils and in the mud at the bottom of lakes and are usually harmless. Type C does not grow below 10°C (Segner, Schmidt & Boltz, 1971) but if high temperatures are accompanied by low oxygen concentrations (such as result from the sudden decay of an algal bloom coupled with low rainfall) the resulting anaerobic conditions are favourable for the proliferation of the bacteria and the production of the toxin. Outbreaks are especially likely in shallow, slightly alkaline lakes with abundant rotting organic matter. The toxin soon builds up to a level which affects waterfowl and remains at lethal concentrations until heavy rainfall dilutes and re-oxygenates the water. Haagsma (1974) found that the level of toxin in lakes in the Netherlands remained undiminished after 9 months. The toxin may be ingested either by drinking or by eating contaminated vegetable or animal matter. Infected birds exhibit progressive weakness and inability to stand or use their limbs. Paralysis of the neck muscles is a characteristic feature (the disease is sometimes known as 'limberneck') and this can lead to drowning, although death is usually due to respiratory or cardiac failure.

The first references to what was clearly botulism describe an outbreak of 'western duck disease' resulting in enormous mortality among waterfowl on Great Salt Lake in the western USA in 1910 (Coale 1911). In many subsequent years there were large numbers of bird deaths, often exceeding 100 000, on lakes and mud flats in the western states, but it was not until 1930 that the cause of death was finally diagnosed as botulism (Kalmbach 1930, Giltner & Couch 1930). Serious outbreaks in Europe are a rather recent phenomenon, with the Netherlands regularly affected since 1970 (Haagsma, Over, Smit & Hoekstra, 1972) and the death of 50 000 birds, probably from botulism, in the Coto Doñana in Spain in August and September 1973.

The first major outbreak in Britain was during the late summer of 1969 when there was a heavy mortality of waterfowl and other birds on the lake in St James' Park, London (Keymer, Smith, Roberts, Heaney & Hibberd, 1972). This 12 acre (*c.* 5 ha) shallow lake usually has a population of something over 1000 birds. The first deaths were observed at the end of June and mortality reached a peak in September, continuing until the middle of November. Over 400 birds of twenty-one species died, mainly ducks and geese but also herring

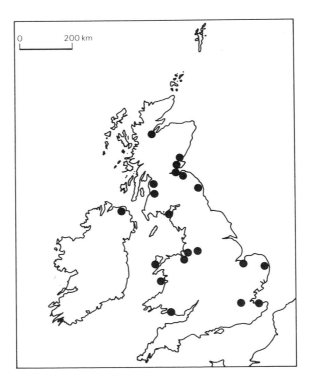

Figure 8.2 Distribution of twenty wild bird mortality incidents attributable to botulism in summer 1975. From Lloyd *et al.* (1976).

gulls (*Larus argentatus*), woodpigeons (*Columba palumbus*), blackbirds (*Turdus merula*), thrushes (*Turdus philomelos*) and starlings (*Sturnus vulgaris*). This outbreak coincided with an unusually long spell of warm weather and there was a heavy bloom of blue-green algae (*Oscillatoria agardhii* and *O. redeki*) on the lake. In the same year there were other suspected outbreaks of botulism including one at Hoveton Great Broad, part of the Bure Marshes National Nature Reserve in the Norfolk Broads. Throughout the 1970s there were various outbreaks in Broadland, mainly centred on Ranworth and adjacent Broads and coinciding with a period of hot dry weather in mid- or late summer. In 1975 the first report of botulism deaths came from Ranworth Broad at the end of June and this outbreak, associated with a particularly warm summer, proved to be the worst yet with over 3000 bird deaths on the Broads. In that year there were at least twenty mortality incidents (Lloyd *et al.* 1976) around the British Isles where botulism was suspected (see Fig. 8.2). There were also outbreaks in the hot summer of 1976, but these seem to have been less serious than those of 1975.

There is a clear connection between outbreaks of botulism and warm, dry climatic conditions which favour the proliferation of the anaerobic *Clostridium* bacteria. Indeed, attempts have been made to establish a predictive relationship between the number of deaths that can be expected and such climatic parameters as the number of days with temperatures above 21°C and the April–July rainfall. However, the recent increase in the number of outbreaks of botulism reflects the general degradation of wetlands such as the Broads. In Broadland, macrophytic vegetation has been replaced by algal blooms largely as a result of phosphate enrichment from sewage outlets. In a warm, dry summer, as the algal blooms die and decay, anoxic conditions are produced and these favour the development of *Clostridium botulinum* so that an outbreak of botulism results.

9

Climatic effects mediated by food

The effects of climate on the distribution and abundance of organisms may result in an indirect response at the next trophic level by those animals for which the directly affected species constitute prey. By such means homeothermic animals such as birds, which are less susceptible to the direct influence of climate, may respond to a climatic change.

9.1 Effect of climate on the distribution of food

If climatic changes bring about changes in the range of a potential prey species (either plant or animal), then one would expect a predator species to respond by altering its own distributional limits in a similar fashion. In fact, there is very little evidence of such responses, mainly due to our lack of knowledge of the diet of many species, the precise distribution of prey species either before or after a climatic change such as the twentieth century warm period, and the ability of predators to modify their diet according to the availability of potential foods.

It is clear that, on a local level, birds, for example, are able to concentrate their search effort in those areas where prey abundance is greatest (Gibb 1962, Royama 1970) and prey distribution will often reflect a response to climate. A familiar example of this relates to the height of feeding swallows (*Hirundo rustica*): on clear anticyclonic days the insects on which they feed are at a high level in the atmosphere and so it is observed that the swallows fly high; on dull, overcast days the insects and the swallows feeding on them are at a lower altitude. On a larger scale, however, there is little evidence of changes of range of land birds in response to climatically associated changes in the distribution of their prey, although the decline and subsequent increase in certain sea bird colonies during the present century may reflect changes in the location of the fish on which they feed in response to changes in sea temperature.

The national census of sea bird colonies around the British Isles ('Operation Seafarer') in 1969–70 (Cramp, Bourne & Saunders 1974) demonstrated the widespread decline in the number of puffins (*Fratercula arctica*) nesting around Britain and especially on St Kilda in the Outer Hebrides (Flegg 1972, Boddington 1960), which formerly had one of the largest colonies in the world. A variety of factors may be involved including soil erosion, an increase in predatory gulls or rats, and the effects of human predation for food and feathers. Flegg found none of these factors to be applicable to the St Kilda colony, where the decline continued after the human population was evacuated in 1930 and where there are no rats and no apparent increase in predatory gulls. As long ago as 1953 it was suggested (Lockley 1953) that climatic changes may have resulted in a reduction in the puffin's food supply, but this view was subsequently discounted (Cramp *et al.* 1974). Recent studies (Harris 1976a) suggest that the decline in the species has been halted and that the more northern colonies are again increasing in numbers. Between 1974 and 1976 the number of occupied burrows on Hirta, St Kilda, increased by 6–7% per year (Harris & Murray 1977) and the bird seems to be increasing again in Iceland after a period of decline. The small colonies at the southern edge of the species' distribution are, however, probably still declining.

Puffin burrows are usually very inaccessible and the bird is consequently very difficult to count. Because of this, our knowledge of the timing of the bird's decline and also the relative severity of the decline at different colonies is inadequate. However, the decline and subsequent increase in the north may be attributed to a change in the distribution of the species' food supply (e.g. species with a northerly distribution such as the sand eel (*Ammodytes* spp.)) as originally suggested by Lockley (1953) and re-asserted by Harris (1976a). According to this hypothesis, with rising sea temperatures between the 1920s and 1950s the fish species on which the puffin feeds changed their location and the puffin colonies declined in consequence. As sea temperatures subsequently declined the cold-water fish returned and the northern colonies of the bird were able to start increasing again. The recent increases in some northern populations of guillemot (*Uria aalge*) (Harris 1976b) may be explained in a similar way, although it should be emphasised that there is little firm evidence for this hypothesis.

The steady increase of the North Atlantic gannet (*Sula bassana*) population through the twentieth century is mainly due to the virtual cessation of its exploitation by man as a food resource. However, there is evidence that the extension of the species' range during this period of climatic amelioration was associated with a climatic effect on the distribution of its food species. New colonies were established in Iceland during the amelioration, apparently associated with an increase in saithe or coal fish (*Pollachius virens*), which, although a northern species, requires temperatures of 6–8°C in order to spawn. Similarly, on the other side of the Atlantic, the re-establishment of colonies in the Gulf of St Lawrence in the 1930s seems to have been a response to an increase in the abundance of mackerel (*Scomber scombrus*) off the Canadian coast.

Herbivorous animals will also respond to the effects of climate on the abundance and distribution of their plant food or 'prey'. For example the perennial marine angiosperm eel grass (*Zostera marina*) is an important food plant of the brent goose (*Branta bernicla*), especially when the birds first arrive on their wintering grounds (Ranwell & Downing 1959). In consequence, the death of the submerged eel grass beds on both sides of the Atlantic in the early 1930s had a serious effect on this species. The American pale-breasted form *B.*

b. hrota declined to less than one-tenth of its former numbers by 1935 (Cottam 1935) and similar declines were noted throughout Europe, including Britain (Salomonsen 1958, Mörzer Bruijns 1955, Atkinson-Willes & Matthews 1960), where the dark-bellied subspecies *B. b. bernicla* occurs. Since the 1950s the numbers of brent geese have been increasing (e.g. Tubbs 1977) as *Zostera* has spread again (it appears that the subspecies *Z.m. angustifolia* is now the dominant form, whereas before the 1930s decline *Z. m. marina* was most abundant) and by 1976 (Ogilvie & St Joseph 1976) the world population of *B. b. bernicla* was about six times what it was in 1955–57 when the first attempt at assessing its size was undertaken.

The high temperatures prevailing in coastal waters during the 1930s have been implicated, either directly or indirectly, in the death of the *Zostera* beds. The decrease has been attributed to disease caused by a bacterium (Fischer-Piette, Heim & Lami 1932), a fungus, e.g. *Ophiobolus halimus* (Mounce & Diehl 1934), or a slime mould (*Labyrinthula macrocystis*) (Renn 1934, 1936), although Young (1938) has shown that the last-mentioned occurs in healthy *Zostera* as well. However, Rasmussen (1973) developed Setchell's (1929) ideas on the importance of a narrow temperature range for the existence of *Zostera* and concluded that the high summer temperatures in the 1930s caused the decline of *Zostera* throughout the northern hemisphere. Although euryhaline, this species is rather stenothermic: metabolic activity ceases below 10°C and between 15 and 20°C the growth of the leaves stops and flowering stems are formed. Above 20°C, however, all activity ceases and temperatures above 25–30°C cause death (Rasmussen 1973). The species rarely flowers in the Mediterranean, probably because of the high temperatures (Benacchio 1938) and temperatures are too low in the Baltic (Luther 1951). An alternative climatic interpretation that the decline was due to fewer hours of sunshine in 1931–32 (Tutin 1938) is untenable (Atkins 1938) and the reversal of the decline when sea temperatures dropped again in the early 1950s provides support for Rasmussen's theory. However, it is still not clear whether the temperature effect is direct or indirect, for example by weakening the plant so that it becomes more susceptible to attack by the pathogens originally suggested as the causative factors.

9.2   Effect of climate on food availability

Apart from the influence of climate on the actual distribution of prey organisms, particular weather factors may render the prey more or less available to the predator at a particular site. A notable example is that of the heron (*Ardea cinerea*) whose population in England and Wales (as evidenced by the number of occupied nests) has been monitored by the British Trust for Ornithology since 1928 (Stafford 1971). The outstanding feature of this long record (see Fig. 9.1) is the essential stability of the heron population at a level of about 4000–4500 pairs (Lack 1954b), but examination of Figure 9.1 shows

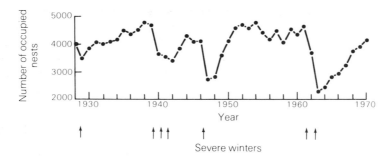

Figure 9.1 Population of heron (*Ardea cinerea*) in England and Wales, as evidenced by number of occupied nests, from 1928 to 1970, and the effects of severe winters. Redrawn from Stafford (1971).

how numbers of nesting herons drop after hard winters, although they recover again rapidly. The cause of the heavy mortality during severe winters (a 30% decline after 1941/2, a more than 50% decline after 1946/7 and a greater than 70% decline after 1962/3) is primarily due to the effect of weather on food availability: the heron is a fish-eating bird and in severe winters many inland ponds and lakes are frozen over for long periods, rendering the fish they contain unavailable to the herons.

The Dartford warbler (*Sylvia undata*) is generally considered to be a year-round resident in Britain (although there is some evidence of a partial migration; Bibby 1979) and in consequence the species suffers high mortality in severe winters. Notable population crashes have occurred following severe weather in 1860/1, 1880/1, 1886/7, 1916/7, 1939/40, 1941/2, 1946/7, 1961/2, 1962/3. In 1960/1 the total British population of *S. undata* probably numbered 460 pairs but the species suffered 80% mortality in the 1961/2 winter and then a 90% mortality (making a total of 98% in two years) in the 1962/3 winter and there were only eleven pairs left by the 1963 breeding season (Bibby & Tubbs 1975). Once again, the climatic influence is expressed via its effect on food availability as prey becomes inaccessible when there is a surface covering of snow on the vegetation. Dartford warblers feed on insects within a size range of 2–5 mm and live mainly on beetles and spiders. Late snowfalls, in February, March or April, have a particularly severe effect and Tubbs (1967) reported a considerable reduction in the number of birds breeding in the New Forest, Hampshire, in 1966 after a late snowfall overnight on 13/14 April and through the following day.

Sometimes snow cover may cause a bird to feed in a different and less optimal habitat. For instance, in winter the Icelandic ptarmigan (*Lagopus mutus*) prefers to feed on *Salix herbacea,* which is the most nutritious of the common shrub species in Iceland (Gardarsson & Moss 1970). In late winter, however, *S. herbacea* becomes inaccessible under thick snow and the bird has to feed on *Betula nana,* which projects out of the snow cover, and on catkins of *B. pubescens,* which are less nutritious.

Extremes of heat can also render food less available to some species, as was evidenced during the British drought of 1976. As earthworms retreated into deeper layers of the soil they became unavailable to ground-feeding birds such as blackbirds (*Turdus merula*) and thrushes (*Turdus philomelos*) as well as the badger (*Meles meles*). On the other hand, the drought reduced the rate of flow of rivers and streams and so increased the availability of fish, including the salmon (*Salmo salar*) to the feral mink (*Mustela vison*), now widespread in Great Britain after escaping from fur farms. This species' voracious habit of surplus killing virtually wiped out the fish in some streams in Devon and the Lake District.

Falling rain may inhibit a predator's hunting ability, as in the case of the tawny owl (*Strix aluco*), which has a lower hunting success in rainy weather (Southern 1970). This is because the tawny owl locates its rodent prey mainly by hearing, which is less accurate when there is a background noise of rainfall. Kingfishers (*Alcedo atthis*) stop diving for prey altogether during periods of rain, presumably because the ripples produced impair visibility.

A similar phenomenon is found in guillemots (*Uria aalge*) which dive for fish such as sprats (*Sprattus sprattus*). Birkhead (1976) investigated how sea conditions affect the rate (number of fish per chick per 4 h) at which guillemots feed their chicks. He found that cloud cover was not an important factor and that there was no effect of wind speed *per se,* but he found a significant negative relationship between feeding rate and sea conditions (calm, moderate and rough) which account for 32% of the variation in feeding rates. The sea surface conditions are not directly related to the prevailing wind speed as a choppy sea persists after the responsible wind has abated. Whilst accepting that, above a certain level, increasing wind speeds and rough surface conditions are detrimental to the fishing success of terns, which plunge-dive from 3–10 m, Dunn (1973) found that such success, expressed as a percentage, and capture rate, i.e. number of fish caught per minute, appear to increase with wind speed in the common tern (*Sterna hirundo*) and the sandwich tern (*Sterna sandvicensis*). Furthermore, for any particular wind speed, moderate sea conditions were generally associated with a higher capture rate than calm conditions. Dunn suggests two possible explanations for these relationships. First, on still days the bird has to make more vigorous efforts in order to hover over the water, and this may make them more conspicuous to the prey; a moderate wind assists a tern in hovering. Second, as the fish is closer to the sea surface than the bird, with an increase in surface disturbance the fish's view of the bird is impaired more than the bird's view of the fish.

The effect of climate on food availability relates to herbivores as well as predators, as may be seen from a number of examples from the Lepidoptera. The small eggar moth (*Eriogaster lanestris*) has decreased dramatically over the past two decades in Britain. The larvae of this species are gregarious and feed on hawthorn (*Crataegus* spp.) and so the destruction of hedgerows may be implicated in the decline of the moth. The adult emerges very early in the year and the eggs are laid on terminal twigs in February and March and hatch soon afterwards. The date of leafing of hawthorn appears to be very variable from year to year, as exemplified by data from the Marsham Phenological Record (see Sec. 6.5), which show that although the average date of leafing was 9 March, the earliest occurrence was on 27 January and the latest was on 26 April (Kington 1974). This suggests that in some years the *E. lanestris* larvae will hatch out before any food in the form of hawthorn leaves is available for them to feed on. The recent tendency to cooler springs (Lyall 1970) may thus account for the decline of the small eggar via its effect on the availability of food to newly hatched caterpillars.

A similar situation is found in the winter moth (*Operophtera brumata*) (Geometridae) which occurs on the oak (*Quercus robur*), sometimes in sufficient numbers to cause complete defoliation. The moths emerge in November and December and pair on the trunks of oak trees. The wingless females then climb up to the branches of the tree and lay their eggs in crevices in the bark. The eggs hatch in April and the larvae pass through five instars before spinning silken threads to the ground where they pupate in late May. The species has been studied in detail at Wytham Wood, Oxfordshire, by Varley and Gradwell (1958, 1963) who concluded that the fluctuations in density from year to year are determined by the density-independent action of weather which influences the relative timing of bud burst on the trees and the insect egg hatch. If the majority of the trees open their buds later than the peak egg hatch, a high mortality of caterpillars results (up to 90%) as the young larvae are unable to penetrate the closed buds. There must be a considerable selective advantage for early feeding in this species because if egg hatch were delayed a few weeks so as to ensure that it always occurred after bud burst, the larvae would be able to utilise the apparent abundance of relatively untouched food in the month of June and after. However, as the season progresses there is a decline in the availability of nitrogen in the oak leaves due to an increase in the concentration of leaf tannins which form complexes with proteins so as to render them unavailable to herbivores (Feeny 1970). For this reason the caterpillars hatch earlier than would appear to be optimal on purely climatic reasons, and in some years this strategy proves unfavourable.

Contrary to the earlier suggestion (Hairston, Smith & Slobodkin 1960) that since plants in most habitats are largely intact and rarely depleted by herbivores then herbivores must have available to them a super-abundance of food, it is now realised (Murdoch 1966, Ehrlich & Birch 1967, Feeny 1970) that much of this apparently abundant food is not available to or not suitable for consumption by herbivores. In fact, many herbivores are probably limited by the quality of their food (particularly its nitrogen content) rather than by its abundance. The nitrogen content of food plants has been implicated in a theory which links locust outbreaks to particular weather conditions (White 1976) which may have a much wider applicability among herbivorous animals. White suggested that changes in the rate of survival of very young hoppers is the major factor influencing changes in the abundance of locusts. He holds the view that there is normally a relative shortage of nitrogen-rich food for the rapidly growing young and that acridid species have adapted to the resulting high mortality by producing large surpluses of young. Under certain weather conditions the food plants become a more nutritious source of nitrogenous food and there is a greatly increased rate of survival of the very young leading to locust outbreaks.

White postulates that changes in the moisture content of the soil, as a result of periods of alternating unusually dry and wet seasons in which the food plants are subjected to periods of drought or waterlogging or both, alter the amount and proportions of nitrogen in the aerial parts of the plants which are eaten by the insects. There is, in fact, considerable evidence that a number of plant species show an increase in the nitrogen (especially amino acid) content of their aerial parts when subject to water stress (e.g. Vaadia, Raney & Hagen 1961, Abdel Rahman 1973, Baskin & Baskin 1974, Kemble & Macpherson 1954, Dove 1968, Singh, Paleg & Aspinall 1973, Hopkins 1968, Barnett & Naylor 1966).

White's hypothesis may prove to be an important mechanism whereby climate exerts an effect on the numbers of herbivorous insects in temperate as well as tropical environments. White has previously shown that outbreaks of the psyllid insect *Cardiaspina densitexta* (White 1969) in Australia and of the geometrid caterpillar *Selidosema suavis* (White 1974) in New Zealand tend to occur in areas where drainage is impeded, making soils prone to seasonal waterlogging and drought, exaggerated by particular weather patterns. Other cases where this mechanism might be involved include outbreaks of the nun moth or black arches (*Lymantria monacha*) in Denmark, which tend to occur in the drier eastern areas of the country after 2–3 years of hot, dry summers (Bejer-Petersen 1972), and outbreaks of the spruce budworm (*Choristoneura fumiferana*), which often occur after several summers of warm, dry weather (Morris 1963, Ives 1974). Similarly, it is interesting that outbreaks of the striped hawk moth (*Celerio lineata*) in America tend to occur when a wet year has followed a dry one (Grant 1937). One is tempted to speculate, for instance, that the high numbers of aphids (whose diet is known to be nitrogen-deficient; see Dixon 1973) present in 1977 in Britain might perhaps be due to an increase in the available nitrogen in plants subject to an extremely wet autumn after a summer drought in 1976. Further research in this field, in particular monitoring year-to-year differences in the nitrogen content of the aerial parts of plants, is urgently needed in order to resolve what might be a very significant effect of climate.

9.3   Effect of climate on prey activity

As discussed in Section 6.7, temperature and other meteorological factors affect the activity of poikilothermic animals. This effect of climate on activity will result in changes in the availability of such animals as prey for organisms higher up the food chain. Those predators which can only detect actively moving prey (such as spiders and frogs) will be particularly affected, as will those which do not actively search for their prey but rely on the prey animals' locomotory activity to bring them within the predator's reactive area (what Schoener 1971 called 'sit-and-wait predators').

As is often the case, there is little direct evidence of such effects operating in the wild; however, Goss-Custard (1969), working on the birds of the Ythan Estuary, Aberdeenshire, in the winters of 1964–66, showed that mud temperature affected the diet and feeding rate of the redshank (*Tringa totanus*), probably because of its effect on the behaviour and thence availability of the prey. The redshank is a small wading bird inhabiting the muddy parts of estuaries and feeding on the rich intertidal macrofauna there. The prey is detected by sight and four species predominate in the redshank's diet: the amphipod crustacean *Corophium volutator* (the most important prey species), the gastropod mollusc *Hydrobia ulvae*, the bivalve mollusc *Macoma balthica* and the polychaete worm *Nereis diversicolor* (the ragworm). Above 6°C, the more active *Corophium* was the main prey species in terms of numbers (87% or more) and biomass (82% or more) (Goss-Custard 1969). As the temperature decreased during winter, the feeding rate of the redshank decreased and, furthermore, their diet changed so that at temperatures from -1 to +1°C sedentary species such as *Macoma* contributed 76% of the biomass ingested at one area and *Nereis* contributed 54% at another, these differences reflecting the different relative densities of the two species at the two sites.

The effect of climate on prey activity has been implicated in the decline of the red-backed shrike (*Lanius collurio*), both in Britain (Peakall 1962, Ash 1970, Bibby 1973) and in northern Europe (Kalela 1949, Durango 1950). The bird has always been more abundant in continental rather than oceanic regions and has, for example, never been reported from Ireland. It is fairly common around Oslo but has never been found along the western coast of Norway, and in Sweden it breeds up to 61°N in the west but extends to 65°N in the drier eastern region. In areas which, although situated away from the Atlantic, still experience high rainfall (e.g. southern Bavaria; Durango 1950) the shrike is also less common.

On the European continent, annual fluctuations in the red-backed shrike seem to occur and in some years the bird is two to three times more numerous. Such increases have been correlated with good weather during migration (Svardson & Durango 1950). However, during the present century, the population of the red-backed shrike has generally been declining in Europe. especially in Sweden, Denmark, Holland, Belgium and north-western Germany, a trend which appears to coincide with a change to a more oceanic climate associated with a more westerly circulation. Over the same period, the shrike withdrew from the more maritime zone in France and in Britain its range contracted towards the south-east (Peakall 1962) so that by 1971 there were only eighty-one pairs left with 76% of this remaining British population (Bibby 1973) being confined to East Anglia, which experiences the most continental climatic regime of any area in the country (Ford & Lamb 1976).

Of course, once a population has been reduced to low levels by the influence of a climatic factor, the downward momentum tends to be maintained irrespective of the prevailing climatic conditions. Fragmentation of the population reduces breeding success and, in the case of the red-backed shrike, the rarity of the bird has increased the value of its eggs to collectors, whose depredations are now a significant factor in maintaining the bird at a low level.

The red-backed shrike feeds largely on moving insects (bees, beetles, grasshoppers, dragonflies, etc.) as well as other active animals, including birds and mammals, and it is likely that the decline in the shrike population reflects the reduced availability of such creatures resulting from their lowered activity in more maritime climatic conditions. Durango (1956) reported experiments with caterpillars, beetles and grasshoppers as 'bait' and concluded: 'When placed in shade the bait soon becomes immobile and does not attract a shrike who may be sitting directly above it. In sunny weather the struggles of the bait are observed by the shrike almost at once.' Durango also records that one territory that he observed was so situated that it was in shadow almost the whole morning with the result that the cock, which feeds the hen as well as the young birds, had to bring food from outside his territory. He found a similar situation on the steep slopes of a mountain at Menton, southern France, where every morning males of *L. collurio* and *L. senator* (the woodchat shrike) from the western slope, then in shadow, moved to the eastern side.

Without reliable information concerning the availability of insects in past decades, it is difficult to find good evidence for the link between the decline of the red-backed shrike and reduced availability of food associated with a climatic change. Bibby (1973) collected population data from published county bird reports and attempted a multiple regression analysis involving sunshine, temperature and rainfall data. Although he failed to find any firm associations, he concluded that dry weather, especially in late summer, is beneficial to the subsequent year's population. Ash (1970), observing a population of shrikes in Hampshire, attempted to categorise the favourability of the months May, June and July for the years 1953–68 but was unable to find any consistent correlations with various aspects of the breeding season.

In 1965, Ash (1970) attempted to assess the abundance of food in different habitats and in different weather conditions by counting the number of large

Table 9.1 Counts of large insects in areas with high and low population levels of red-backed shrikes in good and poor weather conditions (after Ash 1970).

Population level	Weather conditions	Minutes counted	Number of insects	Insects per minute
high	good	150	1010	6.7
high	poor	150	87	0.6
low	good	160	212	1.3
low	poor	35	4	0.1

insects crossing the field of a fixed pair of 10×10 binoculars. One habitat, supporting a large population of *L. collurio,* consisted of heather, gorse and grass in equal proportions whilst the other, supporting a low population of *L. collurio,* comprised equal proportions of heather, gorse and bracken. Ash conducted his insect count in good (temperature above 60°F, sunny, with light winds) and poor (temperature below 60°F, overcast, wet and often windy) weather conditions and obtained the results indicated in Table 9.1. It can be seen that under good weather conditions there were five times more insects in the high population area than in the low, and in both areas there was only about one-tenth the number in poor weather.

The average clutch size of *L. collurio* is five or six eggs laid at a rate of one a day and the incubation period from the time of completion of the clutch is about 14 days (Ash 1970). So for about 3 weeks the cock has to find enough food for himself and the hen and until the young are about a week old he has to provide for them as well, a task which must be difficult in unfavourable weather. Durango (1956) reports that in stormy and rainy weather eggs and small young are destroyed if the hen is forced to leave the nest to hunt for herself. The young usually spend about 14 days in the nest and then are fed for about a month by both the parents. The daily food consumption during the nestling period averages 9.5 g fresh weight per nestling per day or approximately 56% of the mean body weight, whilst the daily food consumption of fledglings averages 14 g fresh weight per bird per day or 50% of mean body weight (Diehl 1971). Diehl (1971) concluded that the amount of food consumed by young birds until a month of age in a red-backed shrike population in Strzelckie Meadows, near Warsaw, Poland, was 3.5 mg fresh weight of insects per square metre per day, to which should be added the food consumption of the adults, which is about 1.7 mg fresh weight per square metre per day.

It is clear then that the energy required to raise a brood of red-backed shrike is considerable and any reduction in prey availability due to adverse weather conditions may have serious results as far as reproductive success is concerned. The effect of weather is further enhanced as the metabolic rate, and hence food consumption per day, of fledgling shrikes is increased at lower temperature (Diehl 1971).

Among other insectivorous birds, as well as bats, it is known that reduced activity of insects in wet summers may result in heavy losses from starvation among the young. This applies to swifts (*Apus apus*) (Lack & Lack 1951) and partridges (*Perdix perdix*) (Potts 1969), although the decline of this latter species in Britain (temporarily arrested by the high breeding success in the exceptionally fine summer of 1976) is mainly associated with a reduction in the availability of insect prey as a result of the destruction of hedgerows in recent years and the spraying of crops with herbicides (Southwood & Cross 1969). Also, during the more oceanic regime earlier in the present century, the insect-eating hoopoe (*Upupa epops*) and the roller (*Coracias garrulus*) became extinct in Denmark (Salomonsen 1948). Durango (1946) investigated the decline of the latter species in Sweden and concluded that, like the red-backed shrike, it decreased in the more westerly parts of its range during the present century due to the effects of the more maritime conditions on the availability of its insect food. At the same time, the species expanded northwards in eastern Europe and Russia, probably in part the result of population pressure due to the relocation of birds formerly inhabiting western Europe. The decline of the white stork (*Ciconia ciconia*) in Sweden, Holland, Belgium, France and Denmark is also probably partly attributable to the effect of climate on the availability of food.

10

Climatic effects mediated
by competition

10.1   Introduction

Competition occurs whenever two individuals, of the same or different
species, interact over the utilisation of a common resource in limited supply.
In the case of interspecific competition, one species will inevitably exploit the
resource more efficiently and will increase numerically at the expense of the
other species. As the balance of competitive advantage changes with changing
environmental, including *inter alia* climatic, factors the outcome of a
competitive interaction may alter with the result that one species may displace
another. This can be demonstrated in controlled experiments such as those of
Park (e.g. Park 1962) on competition in laboratory cultures of flour beetles
living in a completely homogeneous environment. In one set of experiments,
Park (1954) examined the effect of temperature and humidity on the outcome
of competition between two species of the flour beetle *Tribolium*, *T. castaneum*
and *T. confusum*. Although either species could survive under a variety of
climatic conditions provided that it was present alone in the culture, when the
two species were cultured together it was found that *T. castaneum* had an
advantage at high temperatures or humidity while *T. confusum* had an
advantage under conditions of low temperature or humidity. In either case the
disadvantaged species soon became extinct. Similar results are reported by
Birch (1953), who found that with mixed cultures of the beetles *Calandra
oryzae* and *Rhizopertha dominica*, *C. oryzae* had a competitive advantage at
29°C while *R. dominica* had a competitive advantage at 32°C.

Such laboratory experiments suggest that climate may affect the outcome of
competitive interactions in the wild. Such interactions are usually resolved by
spatial separation such as that reported by Beauchamp and Ullyott (1932),
who examined the spatial distribution of two species of triclad planarians or
flatworms (*Planaria montenigrina* and *Planaria gonocephala*) found in
highland springs and streams around Europe. Figure 10.1 demonstrates how
the distribution of the two species is related to temperature in allopatric
(occurring separately) and sympatric (occurring in the same stream)
situations. Beauchamp and Ullyott found that both species occurred alone in
several springs where the water temperature was between 6.5 and 8.5°C, but *P.
montenegrina* can only tolerate temperatures up to 16 or 17°C, while *P.
gonocephala* inhabits waters as warm as 23°C. Where they occur in the same

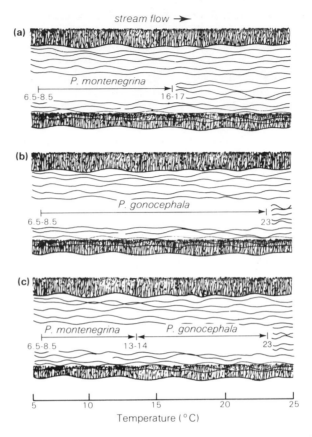

Figure 10.1 Distribution of the flatworms *Planaria montenegrina* and *P. gonocephala* in highland streams in relation to temperature. (a) and (b) represent allopatric situations; (c) represents a sympatric situation.

stream, *P. gonocephala* is excluded by competition from the spring head to a point where the temperature reaches 13–14°C. Downstream from this point *P. montenegrina* is absent and *P. gonocephala* is the only species present.

10.2 Competition in marine environments

If the distributional limits of a species are determined by competitive interactions, then any change in climate is likely to result in a re-adjustment of the species' range. Most of the documented examples of this phenomenon relate to marine environments which are less spatially heterogeneous and in which a single climatic variable, namely temperature, is of over-riding importance.

During the present century there has been a cycle of change in the composition of the marine fauna in the western English Channel which is often referred to as the Russell cycle after its discoverer Sir Frederick Russell (see Russell 1935, 1973, Russell, Southward, Boalch & Butler 1971). This cycle particularly involves the abundance of macroplankton, which in 1930–31 was reduced in abundance by a factor of four, with a similar reduction in summer-spawned fish larvae and in the winter maximum of dissolved inorganic phosphorus.

One of the most striking features of this cycle was that in the early 1930s the chaetognath *Sagitta elegans,* which had formerly characterised the plankton community, was replaced by the related *S. setosa* and the overall abundance of chaetognaths declined (Russell 1935). From about 1965 the plankton community of the western English Channel off Plymouth began once again to resemble, in both composition and abundance, that which predominated before the 1930s and in 1972 *S. elegans* was again found in considerable numbers off Plymouth (Southward 1974) and by the late 1970s the return to the pre-1930s fauna (with the exception of the composition of the pelagic fish fauna) was complete (Southward 1980). The cold water *S. elegans*-dominated plankton community present before the 1930s and returning in the 1970s is also characterised by the trachymedusan coelenterate *Aglantha digitalis* the siphonophore *Nanomia* sp., the polychaete worm *Tomopteris helgolandica,* the euphausid crustaceans *Thysanoessa inermis* and *Meganyctiphanes norvegica* and the pteropod mollusc *Spiratella retroversa.* In addition the starfish *Luidia sarsi* is a cold water indicator which has again been abundant in the western English Channel at certain times in recent years (Southward 1974). During the period when the warm water community characterised by *S. setosa* predominated, *Liriope tetraphyllas* replaced *Aglantha digitalis* and *Nanomia* was replaced by the siphonophore *Muggiaea atlantica.* The euphausid *Nyctiphanes couchi* and the southern ascidian tunicates *Salpa fusiformis* and *Cyclosalpa barberi* (Cushing 1976) also featured in this community and, although calanoid copepods were much less abundant than in the *S. elegans* community, the copepod *Euchaeta couchi* was characteristic.

The changes associated with the Russell cycle appear to coincide with the warming and subsequent cooling of sea temperatures over the period concerned and the rapidity of the replacement of species characteristic of a northerly cold water fauna by southerly warm water species suggests that this response to climate was effected by an influence on the outcome of competitive interactions. This is especially likely to be the case in the two *Sagitta* species where competition between two species of the same genus having very similar ecological requirements would be particularly great. However, as Southward (1980) has pointed out, *S. elegans* was not replaced by the most nearly equivalent warm water species, *S. friderici,* which was only found in small numbers during the warmest phase of the cycle, but by the more neritic (i.e. shallower water species, more characteristic of the continental shelf) *S. setosa,* which is perhaps more opportunistic as it is able to withstand a wider range of temperatures.

Associated with changes in the plankton community in the western English Channel was the replacement of the herring (*Clupea harengus*), a pelagic fish close to its southern limit at the entrance to the Channel, by the more southerly pilchard (*Sardina pilchardus*). The collapse of the commercially important Plymouth herring fishery in the 1930s (Ford 1933) and of the Firth of Forth fishery in the 1940s (Saville 1963) is clearly related to the rise in sea temperatures, but the actual mechanism of the relationship is uncertain. The Plymouth herring fishery failed completely after 1935, when more than 80% of the catch consisted of fish over 6 years old (Kemp 1938). The failure was therefore due to lack of recruits, not lack of adults, and Cushing (1961) suggests that the decline in year-class strength began with fish hatched in 1926 and was completed by 1931. In 1926, pilchard eggs, the numbers of which, incidentally, show signs of an 11 year cycle perhaps associated in some way with sunspot number (Southward, Butler & Pennycuick 1975), were recorded in considerable numbers in July and remained abundant in summertime until 1965–66 (Cushing 1976).

As mentioned above, there was a decline in available winter phosphorus in the Channel in the 1930s and Cushing (1961) showed that the magnitude of the herring recruitment was directly correlated with the quantity of winter phosphorus 1 year after hatching. The quantity of winter phosphorus was inversely correlated with the number of pilchard eggs present 6 months earlier. Thus, herring hatched in winter were inversely related to pilchards hatched the following summer. From this, Cushing concluded that the Plymouth herring fishery failed because the stock became reduced in competition with the pilchard population, and that the change in the winter maximum of phosphorus was the effect of an increase in the number of pilchards – that is, the phosphorus was 'locked up' in the bodies of the pilchards rather than being released in winter as plankton populations decreased. The process was reversed after 1965 (Cushing 1976).

Southward (1963), whilst not disputing the importance of a biological interaction between the herring and the pilchard, concludes that the replacement of the former by the latter can best be explained by a direct response to the higher sea temperatures of the 1920s and later. In his 1961 paper, Cushing makes no mention of the changes of sea temperature over the period under consideration, but in a subsequent review paper (Cushing & Dickson 1976) points out that the two explanations are not exclusive. Thus, the rise in sea temperature altered the balance of competitive advantage between the herring and pilchard via the mechanism elucidated by Cushing (1961).

As with the change in the dominant chaetognath species of the plankton community described above, the replacement of the herring by the pilchard was only indicative of other changes in the fish fauna (Southward 1980): cold water species including cod (*Gadus morhua*), haddock (*Melanogrammus aeglefinus*), ling (*Molva molva*), whiting (*Merlangius merlangus*), plaice (*Pleuronectes platessa*), dab (*Limanda limanda*) and lemon sole (*Microstomus kitt*) declined as warm water species including scad or horse mackerel (*Trachurus trachurus*), sea bream (*Pagellus bogaraveo*) and conger eel (*Conger conger*) have become more abundant. However, in contrast to the situation with the plankton community, the reversal of the cycle in the late 1960s to early 1970s did not produce a return to the *status quo ante* 1930: the pilchard has declined again but the herring has not returned in its former numbers. Instead the mackerel (*Scomber scombrus*) has become the dominant fish and the cold water gadoids Norway pout (*Trisopterus esmarkii*) and blue whiting (*Micromesistius poutassou*) have recently extended into the western English Channel (Southward & Mattacola 1980).

Another classic example of the way in which a climatic change can affect the outcome of a competitive interaction in a marine, or at least intertidal, habitat, and so bring about changes in the distribution of species, relates to the common shore barnacles (Crustacea: Cirripedia) *Balanus balanoides* and *Chthamalus stellatus*. *B. balanoides* is a northern, arcto-boreal species which becomes scarce when water near the shore reaches temperatures above 18°C during the warmest month or above 8.9°C in the coldest month: Hutchins (1947) showed an apparent correspondence between the global distribution of

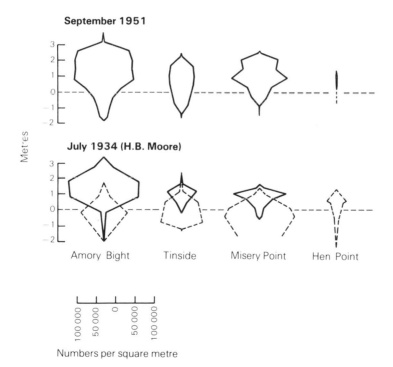

Figure 10.2 Changes in the relative numbers of barnacle species at stations in the Plymouth area: ————, *Chthamalus stellatus*; – – – – –, *Balanas balanoides*. Redrawn from Southward and Crisp (1954).

B. balanoides and the 45°F (=7.2°C) isotherm for winter sea temperature, but the distribution extends south of this isotherm in Europe (Fischer-Piette & Prenant 1956). *Ch. stellatus* is a southern species which reaches its present northern limit in Scotland where the air temperature in the coldest month (February) is about 5°C. The two species only occur together in Britain and northern France, where they reach the southern and northern limits respectively of their distributions.

In 1951 A. J. Southward repeated H. B. Moore's (1936) studies on the zonation of barnacles on shores in the Plymouth area made some 20 years previously. His results, as shown in Figure 10.2, demonstrated that between 1934 and 1951 the northern species *B. balanoides* had virtually disappeared and its place had been taken by the southern species *Ch. stellatus.* Between 1946 and 1952, D. J. Crisp had been making regular observations on the barnacle populations at Brixham, Devon and had found a significant reduction in numbers of both adults and settling spat of *B. balanoides* (Southward & Crisp 1954). In fact, the decrease in *Balanus* populations and the increase in *Chthamalus* numbers was widespread along the western coasts of the British Isles (Southward & Crisp 1956) and was clearly correlated with the rise in sea and air temperatures. After the early 1950s there was a change to cooler conditions resulting in an increase in the *Balanus* populations of southern Devon (Southward & Crisp 1956) and elsewhere on the southern coast of England (Southward 1967), western Ireland (Crisp & Southward 1960) and the Atlantic coast of Europe (Crisp 1965).

In the zone of overlap, the two species compete strongly for space (Southward & Crisp 1956, Crisp 1960) and *Balanus,* which has a higher growth rate, can prise *Chthamalus* larvae off rocks or simply grow over them at lower levels of the shore (Connell 1961). However, the outcome of competition is affected by temperature, via its effect on reproduction and feeding rate. Thus, *Chthamalus* requires temperatures exceeding about 15°C in order to breed (Patel & Crisp 1960) while *Balanus* can only breed after it has been pre-conditioned for several weeks at temperatures below 10°C (Crisp 1957). Similarly, laboratory experiments on the effect of temperature on the rate of beating of the cirri (modified legs), which barnacles use to strain food particles from the surrounding water as well as for respiration, show that *Chthamalus* will not beat its cirri below 5°C and has a maximum beat frequency at 30°C (Southward 1955). *Balanus,* on the other hand, continues to beat its cirri at temperatures down to 1.8°C at least and has a maximum beat frequency at 17°C. In other words, *Chthamalus* beats faster than *Balanus* at temperatures above 17°C whilst *Balanus* beats faster than *Chthamalus* at temperatures below 17°C. This means that wherever they are in competition warmer conditions favour *Chthamalus* and colder conditions favour *Balanus.*

It is interesting that, although it is a characteristically southern species, *Ch. stellatus* was entirely unaffected around most of the British coastline in the severe winter of 1962/3 (in contrast to another southern species, *B. perforatus;* Crisp *et al.* 1964). This indicates that where species are limited by competition, although climatic trends can affect the outcome of this competition, climatic extremes may not of themselves be significant. Where species respond directly to climate, however, it is frequently the individual extreme climatic events which bring about a re-adjustment of the species' distributional limits.

10.3 Competition in terrestrial environments

Examples of climatic changes exerting an influence on the outcome of competitive interactions in terrestrial ecosystems are not well documented mainly because so many other factors, e.g. changes in land use, have an impact in terrestrial situations that it is difficult to distinguish the effects of climate. Even restricting one's consideration solely to climatic factors, the situation is complex as terrestrial organisms respond to the totality of climatic parameters rather than just temperature, which is the predominant variable in marine systems.

However, an interesting example of how climate affects the outcome of competitive interactions among plants is provided by an investigation concerning the distribution of dog's mercury (*Mercurialis perennis*), oxlip (*Primula elatior*) and bluebell (*Endymion non-scriptus*) in Hayley Wood, Cambridgeshire (Martin 1968). Dog's mercury is found on better drained sites within the wood and advances on to poorly drained areas in dry seasons, only to retreat again in wet ones. The plant is susceptible to waterlogging, in which conditions iron compounds in the soil are reduced by the activities of micro-organisms to the soluble ferrous form in sufficient quantities to poison the plant's roots. The oxlip, although capable of surviving on drier sites, is restricted to wetter sites due to competition from the dog's mercury.

Abeywickrama (1949) performed some laboratory experiments to test the relative effects of waterlogging on these two species. One experimental regime was designed to investigate the effects of different heights of water-table for a standard length of time, 70 days. This showed that whilst dog's mercury was killed by a water-table 10 cm or less below the soil surface, oxlip plants could survive for 70 days with the water-table only 5 cm below the ground surface. Another series of tests were performed in which a constant depth of flooding (2 cm) was maintained for different lengths of time. In these experiments it was found that dog's mercury was killed by 18 days submergence in February, but appeared more sensitive later in the season when growth was progressing more actively. Oxlip plants were also rather more susceptible when growing actively, but could generally survive 45 days in 2 cm of floodwater.

The bluebell (*Endymion non-scriptus*) is clearly intermediate in its tolerance of waterlogging and in damper areas dominated by oxlip it is generally found on slightly raised sites whilst in the drier areas where dog's mercury predominates it is restricted to slight hollows. Abeywickrama's experiments on this species were rather inconclusive as tolerance to water-table level, and also flooding, varied with the age of the bulb. However, mature bulbs could withstand 2 cm of water above the soil for 24 days, which is a time intermediate between that obtained for dog's mercury (18 days) and oxlip (45 days).

Detailed investigations of this sort into the tolerance limits of plant and animal species would probably produce further examples of climate affecting the distribution of organisms by means of competitive exclusion, as in the altitudinal distribution of the rose-root (*Sedum rosea*) and the orpine (*Sedum telephium*) (Crassulaceae) (Woodward & Pigott 1975, Woodward 1975). *S. rosea* is an arctic–alpine species with a lower altitudinal limit of 400 m in England, while *S. telephium* is a continental species which reaches its upper altitudinal limit of 400 m in the Lake District. The growth of *S. rosea* appears to be insensitive to altitude but at lower altitudes *S. telephium* grows taller and larger than *S. rosea* and is dominant. At higher altitudes growth of *S. telephium* declines, as does flower production and seed germination, so that above 400 m *S. rosea* is the dominant species. Growth experiments (Woodward 1975) demonstrated that the change in the outcome of competition between the two species at different altitudes is due to the sensitivity of *S. telephium* to temperature.

Finally, before leaving the plant kingdom, there is a very interesting example of a climatic effect mediated by both competition and pathogen attack from Steep Holm, an island situated in the Severn Estuary at 51°21′N 3°07′W (J. Fowles, personal communication). This island is the sole British site of the rare wild peony (*Paeonia mascula*) and the surviving population is currently endangered due to competition from the dominant plant alexanders (*Smyrnium olusatrum*) (the wild peony was introduced from southern Europe in the twelfth century, probably by the monks of the priory on Steep Holm who used its roots and seeds for pharmaceutical purposes). The seed of alexanders germinates in autumn and in mild wet winters, such as predominated in the early 1970s, produces a luxurious 'forest' which shades the wild peony and renders it more susceptible to a botrytal infection.

Analogous terrestrial examples from the animal kingdom are not well documented, although the northwards spread of the curlew (*Numenius arquata*) in Britain and northern Europe during the present century at the same time as the closely related whimbrel (*Numenius phaeopus*) has been retreating in the same direction has been interpreted as an example of a climatic change influencing the outcome of a competitive interaction. The fact that the whimbrel has been increasing and spreading south again in recent years suggests that the trend to cooler conditions is altering the balance of competitive advantage between these two species.

However, the difficulty of attributing such changes in range to the operation of climatic factors rather than other environmental influences is exemplified by a re-examination (Järvinen & Väisänen 1979) of the work of Merikallio (1951) on the influence of twentieth century climatic changes on the competitive interactions between two pairs of congeneric bird species whose ranges overlap in Finland. The species concerned were the northern

Siberian tit (*Parus cinctus*) and the southern crested tit (*Parus cristatus*) and the northern brambling (*Fringilla montifringilla*) and the southern chaffinch (*Fringilla coelebs*). Although the changes in distribution of these species – *P. cinctus* initially retreated northwards as *P. cristatus* expanded polewards but then the balance of advantage shifted as *P. cristatus* declined again in the north; *F. coelebs* increased during the present century although *F. montifringilla* only decreased temporarily in the 1940s – can readily be interpreted in terms of what is known of twentieth century climatic changes, Järvinen and Väisänen (1979) have shown that habitat changes, especially the expansion of forestry, provide a more consistent explanation of the changing distribution of these bird species.

11

Résumé – the levels of biological response to climatic change

Against a background of, on the one hand, the basic biological principles involved (Ch. 2) and, on the other, an account of the changing climate (Chs 3 and 4) and the way that it interacts with environmental features (Ch. 5), the subsequent chapters of this book have been concerned with the variety of ways in which the natural fauna and flora respond to climatic change. It is now clear that there are a number of levels of response, ranging from the response of individual organisms to that of whole ecosystems.

11.1 Responses of the individual and of the species

Individual organisms may exhibit metabolic or physiological responses to climatic variations (Ch. 6, see also Ch. 2), provided that such variations are of restricted amplitude within boundaries defining the organism's tolerance limits. Organisms inhabiting relatively unvarying environments exhibit a narrow tolerance range whilst those inhabiting environments with, for example, considerable annual variations of climate, exhibit a much wider range of environmental tolerance. The former are termed stenotopic and are clearly more vulnerable to unpredictable environmental change than the latter, which are termed eurytopic. Similar terms may be used in respect of particular environmental factors, e.g. stenothermous or eurythermous meaning unable or able to tolerate a wide variation in temperature and stenohaline or euryhaline meaning unable or able to tolerate a wide variation in osmotic pressure.

As described in Section 2.1, tolerance limits are related to the environmental conditions, in general the temperature, at which the enzyme complex of an organism functions optimally, and so they are, ultimately, genetically determined. However, as tolerance limits have evolved in adaptation to the prevailing temperatures in an organism's habitat, they are not necessarily constant within a species; thus populations occurring in more oceanic areas are frequently more stenotopic than those subject to the wider environmental variation characteristic of a more continental regime (Sec. 2.5). Furthermore, within such genetically determined limits, if exposed to a gradual environmental change an individual organism is often able to adjust its physiological tolerance by a process of slow adaptation of the preferred optimum of its enzyme-based physiological reactions, such a process being known as acclimatisation. There is a considerable literature (see, for example, Bullock 1955, Anderson 1970, Wieser 1973) on experimental manipulation of environmental temperature demonstrating the rapidity of such adaptation of physiological processes like respiration, and in predictably variable environments such as temperate regions such acclimatisation occurs from season to season. By extension, long-lived individual organisms may similarly be able to adapt to inter-annual environmental changes and so remain closely adapted to changing climatic conditions.

Over the longer term, however, genetic adaptation of enzymically determined tolerance limits will be a more important response, in terms of evolution, of a species to climatic change. As evolutionary success is determined by the genetic contribution that an individual makes to the next generation, genetic adaptation of tolerance limits will proceed most rapidly in short-lived species whose gene complexes are repeatedly re-presented for the action of selection pressures.

It is characteristic of such a genetic response to climatic change that it is not manifest to the observer unless the organism in question is subject to experimental manipulation in order to define its tolerance limits. There are, however, other genetic responses to climatic changes, the adaptive significance of which are more apparent; for example the thermal advantage enjoyed by dark-coloured forms of certain poikilotherms may contribute to the selective advantage of melanic morphs (see Sec. 6.6) and the various banding morphs of the shell of snails of the genus *Cepaea* are correlated with climatic factors, although the selective advantage associated with these correlations (e.g. the bandless morph of *C. hortensis* being predominant in cold, humid areas) is not always obvious.

A similar genetic response to climate, the underlying mechanism of which is not fully understood, is to be found in the mottled grasshopper (*Myrmeleotettix maculatus*), which occurs in Britain mainly in dry, sandy environments such as heaths and sand dunes. Like a number of other grasshopper species, some British populations of this animal possess supernumerary chromosomes, termed B-chromosomes, which are not found in any of the continental European populations (Hewitt & John 1970). The frequency of these B-chromosomes is correlated with sunny, dry conditions and so southern populations of the grasshopper, inhabiting more favourable habitats, possess supernumerary chromosomes whilst they are absent from more northern populations. There are, however, some areas in southern Britain where the populations do not possess B-chromosomes, one of these being near Aberystwyth, North Wales. Here the frequency of B-chromosomes is negatively correlated with rainfall: the ecologically marginal populations in the West Plynlimon Mountains do not possess B-chromosomes but the frequency of supernumeraries increases as one proceeds either eastward or westward into areas of lower rainfall (Hewitt & John 1967). In addition, some populations in East Anglia, a supposedly optimal area with uniformly low rainfall and high insolation, do not have B-chromosomes: populations in the north and east possess them, but not those in the south and west. In this case the incidence of supernumeraries is correlated with summer temperatures which are lower to the north and east due to the cooling effect of shore winds from the North Sea. This maritime effect makes the localities near the coast ecologically more marginal for this species than the central ones which enjoy a more continental climate (Hewitt & Brown 1970). The basis for this climatic correlation has yet to be fully elucidated, but Hewitt and Brown (1970) have suggested that the presence of a B-chromosome may slow down the rate of development of the grasshopper and this will be a selective disadvantage at the cooler margins of the species' range in which locations the so far undefined selective advantage of the possession of B-chromosomes will be outweighed.

In general, the main manifestation of the response of a particular species to a climatic change is an adjustment of the organism's distributional range. In the case of a climatic deterioration, such a response may be very rapid, for example if it leads to a transgression of the organism's tolerance limits so that death *in situ* results. This will occur in the case of rooted plants and small poikilotherms which are unable to move away from the extreme conditions. A less immediate, but still rapid, response may occur within a single season if the climatic deterioration results in the slowing down of development so that the organism is not able to reach maturity before the growing season ends. In the case of animals this may result in death unless, as with, for example, some butterflies which can overwinter as hibernating larvae or pupae, the organism is able to persist in an inert state so as to complete development in the subsequent season. Of course, as described in Sections 6.2 and 6.3, the climatic conditions required for successful reproduction are on occasion more stringent than those required for normal growth and it may be that, although development is re-commenced in subsequent years, reproduction is never again attainable unless the climate improves again. This means that the species' range will ultimately contract, the timing of this event being determined by the duration of the lifespan of the surviving non-reproductive individuals. As demonstrated in Section 6.2, the ranges of some species of plant are extremely intransitive to changes in climate because they are able to persist vegetatively at a site for hundreds of years.

The speed of response of a species' distributional range to a climatic amelioration is primarily dependent upon the organism's ability to advance into areas previously beyond its tolerance limits, i.e. its mobility. Clearly, animals are generally more mobile than plants and so their distributional limits manifest a more rapid response time. In contrast, terrestrial plants are usually only mobile at one stage of their life-cycle: when their seeds or other propagules may be dispersed by wind or ocean currents. Clearly the smaller and lighter the seed, the greater the species' facility for rapid extension of range. In fact the seeds of many species of plant are continuously being carried beyond the species' normal range and these seeds will only germinate in occasional favourable years. Unless such favourable years presage a general amelioration of the climate, the species' distribution will not be permanently affected.

Longer lived plants such as trees tend to have seeds which are larger and consequently less mobile. Furthermore, a plant species will extend its range in a series of waves or leaps, the distance accomplished in each leap being determined by the mobility of the species' propagules and the time between leaps being determined by the generation time of the plant, i.e. the time taken for the pioneer individuals to develop to maturity and themselves produce seeds which can accomplish the next advance in range. Thus trees, with their long generation time, will once again be rather slow in taking advantage of a climatic amelioration (although, by the same token, tree-lines will be slow in registering a climatic deterioration as the individual trees will persist vegetatively for their alotted lifespan even though no longer able to reproduce). For these reasons the past history of climatic change as determined by the pollen record (that is, the record of the relative abundance of the pollen of those tree species dominant at different times) may not register short periods of warmth such as those around 43 and 13.5 ka BP indicated by the more rapidly responding beetle fauna (Coope *et al.* 1971; see also Sec. 4.2).

Turning now to animals, it is clear that as far as independent mobility is concerned, larger species, being generally more active, can accomplish greater distances and so should respond to climatic changes by more rapid alterations of their distribution. Thus a terrestrial mammal, such as the polecat (*Mustela putorius*), which advanced some 200–300 km north in Finland in the early decades of the twentieth century (Kalela 1949), is able to take advantage of a climatic amelioration by range extension much more rapidly than, for example, a non-flying invertebrate. However, as discussed in Chapter 7, the situation is complicated by the fact that flying animals may be assisted by wind transport just as are the seeds of many plants. Thus aerial animals (mainly birds and insects) may respond more rapidly than purely terrestrial ones. Similarly, marine animals and plants may rapidly extend their range with the assistance of ocean currents (Section 7.1). Migratory species are clearly most susceptible to the influence of wind-assisted transport on the occasions when they make their seasonal long-distance journeys, but the significance of such assistance in extending the species' range depends upon whether the prevailing wind is in the appropriate direction and is thus essentially a matter of chance – see, for example, the northward displacement of Scandinavian fieldfares (*Turdus pilaris*) attempting to migrate south towards Britain, which resulted in the extension of the species' breeding range to Greenland (Salomonsen 1951; see Section 7.4).

11.2   Responses involving interactions between two species

As discussed in the previous section, the primary way in which animal and plant species manifest a *direct* response to climatic changes is by adjusting their distributional limits. However, earlier chapters have demonstrated the manner in which climatic changes can affect the outcome of interactions between species, often resulting *indirectly* in changes of range. The outcome of all the main ecological interactions, predator–prey relationships (Ch. 9), pathogen–host relationships (Ch. 8) and interspecific competition (Ch. 10) are susceptible to the influence of climate.

One of the main ways in which climate influences the outcome of predator–prey interactions is via the direct response of the prey organism resulting in a change in its distribution, such as described in the previous section. This direct response of the prey species is then translated up the food chain. In the light of the great variety of factors affecting the outcome of predator–prey relationships, the most clear-cut examples of the over-riding importance of the prevailing climatic factor, i.e. temperature, come, as so often, from the relatively homogeneous marine environment; see for example the changes in the distribution of sea birds in response to changes in the location of their fish prey discussed in Section 9.1.

As well as influencing the outcome of predator–prey interactions by affecting the location of prey, climatic influences may affect the quantity of prey available to a predator, either by directly influencing the prey population number or else its accessibility. In this context accessibility may relate to the presence or absence of a climatically induced physical barrier such as the way that ice on ponds reduces the accessibility of fish prey to herons and snow cover on the ground reduces the accessibility of terrestrial invertebrates to the Dartford warbler (see Sec. 9.2), or it may relate to the influence of climatic factors on the activity of prey animals and hence their apparent or relative accessibility, as in the case of the red-backed shrike discussed in Section 9.3.

By analogy, in the case of herbivores we can substitute the term 'food' for the term 'prey' and once again we can distinguish the effect of climate on food quantity, for example the climate-related decline in *Zostera* beds and its effect on the brent goose (*Branta bernicla*) (Sec. 9.1), from its effect on accessibility, as exemplified by the effect of the climatic influence of the timing of the opening of buds or leafing of trees on the populations of invertebrate herbivores such as the winter moth (*Operophtera brumata*) (Sec. 9.2). Furthermore, in the case of herbivorous animals which are, in general, utilising food of sub-optimal nitrogen content, climate may exert an influence on food quality by affecting the availability of nitrogen-rich amino acids (see Sec. 9.2).

Finally, climatic influences may affect the duration of time over which the individuals of poikilothermic species may be available as prey to a predator because, as explained in Section 6.3, at lower temperatures the developmental rate of poikilotherms is slowed up with the result that the prey is consistently available for a longer period of time – see for example the increased impact of bird predation on the population of the black hairstreak butterfly (*Strymonidia pruni*) in years with cool springs resulting in an increased duration of susceptible developmental stages (Thomas 1976).

Turning to pathogen–host relationships, it is clear that the influence of climate on these types of interaction is in many ways similar to the climatic response of predator–prey relationships, although the analogy is complicated by the fact that in the former situation the host, i.e. the organism being 'preyed' upon, is several orders of magnitude larger than the attacking organism, the pathogen. In consequence it is not always the case that, as in predator–prey relationships, the direct climatic response is manifest by the organism lower in the food chain and the next trophic level responds indirectly. Thus, whilst a predator may respond indirectly to climate-mediated changes in the distribution of its prey, in the case of pathogen–host interactions it is the smaller pathogen whose distribution may be directly affected by climatic influences with a consequent indirect response by the host species. On occasion there may be a dual response to climate with the direct response of a pathogen resulting in an indirect reduction in the population of its host (as may have happened with the decline of eel grass (*Zostera*) in the 1930s) which is then translated up a predator-prey food chain (in this example the response being a decline in the brent goose (*Branta bernicla*) which feeds on eel grass – see Section 9.1).

Likewise, the direct effect of climate on prey activity, resulting in changes in the accessibility of the prey to the predator discussed above, has a similar 'inverted' analogy in those pathogen–host relationships involving vector transmission (see Sec. 8.2) where the effect of climate on vector activity may have profound effects on the rate of spread of disease. Of course, there is also the possibility of climate directly influencing pathogen dispersal by means of the transport effect discussed in Chapter 7, resulting for example in the cross-Channel dispersal of foot-and-mouth virus over 150 miles (240 km) from northern France to England in March 1981 on the prevailing southerly winds.

Finally, the way in which the slowing of developmental rate increases the period of time over which a prey organism is susceptible to predation has a direct analogy in pathogen–host interactions where a similar developmental response may result in plants and poikilothermic animal host organisms being vulnerable to pathogen attack for a longer period of time.

Let us turn now to the influence of climatic factors on the outcome of competitive interactions. As discussed in Section 2.4, the fundamental ecological niche of a species is essentially defined by the species' physiological tolerance limits as the potential niche that the species could occupy in the absence of any interspecific competition for resources. Across this broadly defined tolerance range, a species is able to exploit the resources of its environment with varying efficiency. When the fundamental niches of two species overlap the resulting competitive interaction is resolved by the partition of the shared resource in a manner determined by the relative efficiency of its exploitation by the two species, with the result that they occupy mutually exclusive realised niches. Where exactly the boundary is drawn, i.e. the balance of competitive advantage, varies with the prevailing environmental, including climatic, conditions.

As the climate changes, the balance of competitive advantage shifts and an adjustment of the distributional boundaries of the competing species ensues. Thus, as sea temperatures around the coast of Britain increased during the present century, the herring (*Clupea harengus*) fisheries were replaced, first off the coast of Plymouth in the 1930s and then a decade later near the Firth of Forth, by the pilchard (*Sardina pilchardus*) (see Sec. 10.2), the latter species apparently being more efficient at exploiting the available environmental phosphorus at higher temperatures (Cushing 1961). Also around the coast of Britain, over broadly the same period, the barnacle *Chthamalus stellatus* increased at the expense of the species *Balanus balanoides* whose feeding efficiency declines at higher temperatures (see Sec. 10.2).

11.3   Responses of communities and ecosystems

It is extremely difficult to evaluate the influence of climatic factors on biological interactions involving more than two species. On occasion, responses at the community level become apparent, although the mechanisms involved remain unresolved. For example it subsequently became apparent that the classic case of the influence of climatic change on the competitive interaction between two species of planktonic chaetognath in the English Channel in the 1930s, which resulted in the replacement of the cold water species *Sagitta elegans* by *S. setosa,* a form tolerant of a wider range of temperatures, was merely indicative of a change in the whole plankton community, as described in Section 10.2.

In the case of terrestrial communities, which are subject to the influence of a much greater variety of natural environmental variables than are communities in marine ecosystems, responses to recent climatic changes are difficult to disentangle from all the other environmental changes that are occurring, in particular those attributable to man's influence.

However, on the broader scale it is clear that the latitudinal distribution of the major terrestrial ecosystems (see Fig. 11.1), paralleled by an analogous altitudinal arrangement up mountains reflecting the fact that temperature drops 0.6°C per 100 m, is climatically determined. These systems are usually called biomes and are characterised in terms of the dominant vegetation which has developed in adaptation to the prevailing climatic conditions; for example the treeless tundra biome, boreal coniferous forest or taiga biome, deciduous forest biome, temperate grassland biome, chaparral or maquis (i.e. Mediterranean vegetation) biome, tropical rain forest biome and savanna grassland biome. The biological record of past climatic changes (see Sec. 4.2) indicates that these biomes have migrated equatorwards and polewards in line with periods of cooling and warming and similar changes, although on a smaller scale, are still apparent where extreme climatic conditions prevail along the northern and southern fringes of the Sahel region of Africa (see Sec. 3.3).

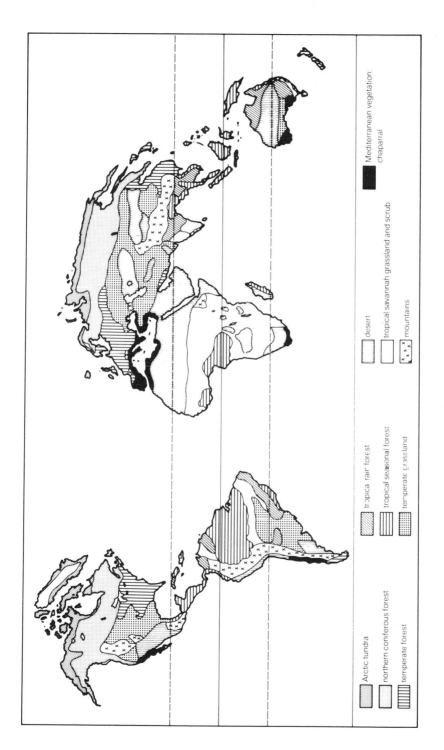

Figure 11.1 Major terrestrial biomes of the world. Redrawn from Cox and Moore (1980).

As the ice retreated after the last glaciation, the grass– and sedge-dominated tundra ecosystem left behind in its wake was gradually transformed into something equivalent to the taiga biome with conifers and birches becoming dominant during the Boreal Period of 9000–8000 BP. As species with higher temperature requirements spread north, a temperate deciduous forest biome was established as the natural climax vegetation of Britain and large areas of equivalent latitude in north-western Europe and north-eastern America. Within this dominant vegetation type there have been changes in the relative abundance of particular species in line with changes in prevailing climatic conditions (see Sec. 4.4). Cold-adapted birch (*Betula*) and poplar (*Populus*) were replaced by hazel (*Corylus*), elm (*Ulmus*) and oak (*Quercus*), with alder (*Alnus*) increasing in wetter times, and then the warmth-loving lime (*Tilia*) became dominant in the Postglacial Climatic Optimum of about 7000–5000 BP. However, from this latter date it becomes more and more difficult to disentangle the climate-related responses of natural communities from effects brought about by the influence of man, a problem exemplified by the various interpretations of the elm decline around 5000 BP which is discussed in Section 4.4.

From this period onwards man's influence on the natural climax communities, especially in respect of their conversion for agricultural production, became predominant; at the present time such systems only survive in the developed world as a series of islands acting as refuges for the natural fauna and flora. If these reservoirs are not climatically optimal for the species that they contain (and most of them only persist because they are considered to be marginal for cultivation either by virtue of their climate or the nutritional status of their soil), then a climatic change (in particular an increase in the frequency of climatic extremes) may have marked deleterious effects – as in the extinction of the subspecies *masseyi* of the silver studded blue butterfly (*Plebejus argus*) (Sec. 5.4), the elimination of the Dartford warbler (*Sylvia undata*) from the remaining islands of suitable habitat in Surrey and north-eastern Hampshire (Sec. 5.4) and the possible future demise of the isolated Lancashire population of the sand lizard (*Lacerta agilis*) (Sec. 2.2). Such population extinctions have occurred continuously in the past, but the continuity of the habitat in former times generally permitted species to repeatedly re-colonise areas from climatically favourable refuges. The rate of man-induced destruction and fragmentation of natural habitats is continuously increasing and, now that it appears that the climatically equable conditions that prevailed in the early decades of the present century are over, it is important to ensure that sites set aside for the conservation of the natural fauna and flora are designated with due regard to the species' climatic requirements and how these interact with features of the habitat, as discussed in Chapter 5. It is probably not possible to influence directly the changing climate, but it may be possible to mitigate the deleterious effects of a climatic change on the natural fauna and flora by conserving them in climatically optimal habitats.

References

Abdel Rahman, A. A. 1973. Effect of moisture stress on plants. *Phyton* **15** 67–86.

Abeywickrama, B. A. 1949. *A study of the variations in the field layer vegetation of two Cambridgeshire woods.* PhD thesis, Cambridge University.

Aguesse, P. 1959. Notes biologiques sur l'éclosion des oeufs de quelques Libellulidae. *Terre et Vie* **1959**, 165–73.

Alexandre, P. 1976. *Le climat au Moyen Age en Belgique et dans les régions voisines (Rhenanie, Nord de la France).* Centre Belge d'Histoire Rurale Publication No. 50, Liège.

Alexandre, P. 1977. Les variations climatiques au moyen age (Belgique, Rhenanie, Nord de la France). *Ann.: Econ. Soc., Civil.* **32**, 183–97.

Anderson, J. F. 1970. Metabolic rates of spiders. *Comp. Biochem. Physiol.* **33**, 51–72.

Andersson, G. 1902. Hasseln i Sverige fordom och nu. *Sveriges Geol. Undersökning Afhand. Serie C2* No. 3, Stockholm.

Andrewartha, H. G. and L. C. Birch 1954. *The distribution and abundance of animals.* Chicago: University of Chicago Press.

Arakawa, H. 1954. Fujiwhara on five centuries of freezing dates of Lake Suwa in Central Japan. *Arch. Met. Geophys. Bioklimatol. B* **6**, 152–66.

Arakawa, H. 1955. Twelve centuries of blooming dates of the cherry blossoms of the city of Kyoto and its own vicinity. *Geofis. pura appl.* **30**, 147–50.

Arnason, E. and P. R. Grant 1976. Climatic selection in *Cepaea hortensis* at the northern limit of its range in Iceland. *Evolution* **30**, 499–508.

Ash, J. S. 1970. Observations on a decreasing population of red-backed shrikes. *Br. Birds* **63**, 185–205, 225–39.

Aston, R. J. 1968. The effect of temperature on the life cycle, growth and fecundity of *Branchiura sowerbyi* (Oligochaeta: Tubificidae). *J. Zool. Lond.* **154**, 29–40.

Atkins, W. R. G. 1938. The disappearance of *Zostera marina*. *J. Mar. Biol. Assoc. UK* **23**, 207–10.

Atkinson-Willes, G. L. and G. V. T. Matthews 1960. The past status of the brent goose. *Br. Birds* **53**, 352–7.

Bakker, R. T. 1971. Dinosaur physiology and the origin of mammals. *Evolution* **25**, 636–58.

Bakker, R. T. 1972. Anatomical and ecological evidence of endothermy in dinosaurs. *Nature, Lond.* **238**, 81–5.

Balfour, E. 1972. *Orkney birds: status and guide.* Stromness: Senior.

Bantock, C. R. 1974. *Cepaea nemoralis* (L.) on Steep Holm. *Proc. Malac. Soc. Lond.* **41**, 223–32.

Bantock, C. R. 1980. Variation in the distribution and fitness of the brown morph of *Cepaea nemoralis* (L.) *Biol J. Linn. Soc.* **13**, 47–64.

Bantock, C. R. and D. J. Price 1975. Marginal populations of *Cepaea nemoralis* (L.) on the Brendon Hills, England. I. Ecology and ecogenetics. *Evolution* **29**, 267–77.

Barica, J. and J. A. Mathias 1979. Oxygen depletion and winterkill risk in small prairie lakes under extended ice cover. *J. Fish. Res. Bd Can.* **36**, 980–6.

Barnett, N. M. and A. W. Naylor 1966. Amino acid and protein metabolism in Bermuda grass during water stress. *Pl. Physiol., Lancaster* **41**, 1222–30.

Baskin, C. C. and J. M. Baskin 1974. Responses of *Astragalus tennesseensis* to drought. Changes in free amino acids and amides during water stress and possible ecological significance. *Oecologia, Berl.* **17**, 11–16.

Beatson, S. H. and J. S. Dripps 1972. Long-term survival of cockroaches out of doors. *Environ. Health* **80**, 340–1.

Beauchamp, R. S. A. and P. Ullyott 1932. Competitive relationships between certain species of fresh-water triclads. *J. Ecol.* **20**, 200–8.

Beebee, T. J. C. 1974. *The natterjack toad* (Bufo calamita) *in the British Isles: a status report.* London: Conservation Committee of the British Herpetological Society.

Beebee, T. J. C. 1976. The natterjack toad (*Bufo calamita*) in the British Isles; a study of past and present status. *Br. J. Herpetol.* **5**, 515–21.

Beebee, T. J. C. 1979. A review of scientific information pertaining to the natterjack toad *Bufo calamita* throughout its geographical range. *Biol Conserv.* **16**, 107–34.

Beirne, B. P. 1947. The seasonal abundance of the British Lepidoptera. *Entomologist* **80**, 1–5.

Bejer-Peterson, B. 1972. Nonnen, *Lymantria monacha* L., i Danmark. (Lep., Lymantridae). *Ent. Meddr.* **40**, 129–39.

Bell, B. 1970. The oldest records of the Nile floods. *Geogr. J.* **136**, 569–73.

Bell, J. N. B. and J. H. Tallis 1973. Biological Flora of the British Isles: *Empetrum nigrum* L. *J. Ecol.* **61**, 289–305.

Bell, W. and A. Ogilvie 1978. Weather compilations as a source of data for the reconstruction of European climates during the Mediaeval period. *Climatic Change* **1**, 331–48.

Benacchio, N. 1938. Osservazioni sistematiche e biologiche sulle Zosteracee dell'Alto Adriatico. *Thalassia* **3**, 1–39.

Benham, B. R. 1973. The decline (and fall?) of the large blue butterfly, *Bull, Amat. Ent. Soc.* **32**, 88–94.

Benham, B. R., D. Lonsdale and J. Muggleton 1974. Is polymorphism in two-spot ladybirds an example of non-industrial melanism? *Nature, Lond.* **249**, 179–80.

Bergthórsson, P. 1969. An estimate of drift ice and temperature in Iceland in 1000 years. *Jøkull* **19**, 94–101.

Berthet, P. 1964. *L'activité des Oribatides (Acari: Oribatei) d'une chênaie.* Mem. Inst. R. Soc. Nat. Belg. No. 152.

Berthold, P. 1973. On the marked decline of the whitethroat *Sylvia communis* and other species of song-birds in western Europe. *J. Orn., Lpz* **114**, 348–60.

Betin, V. V. and J. V. Preobazensky 1959. Variations in the state of the ice on the Baltic Sea and in the Danish Sound (in Russian). *Trudy Gas. Okean. Inst.* **37**, 3–13.

Beverton, R. J. H. and A. J. Lee 1965. Hydrographic fluctuations in the North Atlantic Ocean and some biological consequences. In *The biological significance of climatic changes in Britain,* C. J. Johnson and L. P. Smith (eds), 79–107. London: Academic Press.

Bibby, C. J. 1973. The red-backed shrike: a vanishing species. *Bird Study* **20**, 103–10.

Bibby, C. J. 1979. Mortality and movements of Dartford warblers in England. *Br. Birds* **72**, 10–22.

Bibby, C. J. and C. R. Tubbs 1975. Status, habitats and conservation of the Dartford warbler in England. *Br. Birds* **68**, 177–95.

Birch, L. C. 1953. Experimental background to the study of the distribution and abundance of insects. III. The relation between innate capacity for increase and survival of different species of beetles living together on the same food. *Evolution* **7**, 136–144.

Birkhead, T. R. 1976. Effects of sea conditions on rates at which guillemots feed chicks. *Br. Birds* **69**, 490–2.

Bishai, H. M. 1960. The effects of water currents on the survival and distribution of fish larvae. *J. Cons. Perm. Int. Explor. Mer* **25**, 134–46.

Blus, L. J., R. M. Prouty, and B. S. Neely Jr 1979. Relation of environmental factors to breeding status of royal and sandwich terns in South Carolina, USA. *Biol Conserv.* **16**, 301–20.

Boddington, D. 1960. Unusual mortality of young puffins on St Kilda, 1959. *Scott. Birds* **1**, 218–220.

Boyd, H. 1954. The 'wreck' of Leach's petrel in the autumn of 1952. *Br. Birds* **47**, 137–63.

Brandt, W. 1980. *North-south: a programme for survival.* Report of the Independent Commission on International Development Issues under the Chairmanship of Willy Brandt. London: Pan Books.

Bretherton, R. F. 1951. The early history of the swallow-tail butterfly (*Papilio machaon*) in England. *Entomologist's Rec. J. Var.* **63**, 206–11.

Broecker, W. S. and P. C. Orr 1958. Radiocarbon chronology of Lake Lahontan and Lake Bonneville. *Geol Soc. Am. Bull.* **69**, 1009–32.

Broecker, W. S., M. Ewing and B. C. Heezen 1960. Evidence for an abrupt change in climate close to 11 000 years ago. *Am. J. Sci.* **258**, 429–48.

Brooker, M. P., D. L. Morris and R. J. Hemsworth 1977. Mass mortalities of adult salmon (*Salmo salar*) in the R. Wye, 1976. *J. Appl. Ecol.* **14**, 409–17.

Brown, G. M. 1974. A new solar–terrestrial relationship. *Nature, Lond.* **251**, 592–4.

Brown, L. R. 1979. *Resource trends and population policy: a time for reassessment.* Worldwatch Paper No. 29. Washington, DC: Worldwatch Institute.

Bryson, R. A. and B. M. Goodman 1980. Volcanic activity and climatic changes. *Science, NY* **207**, 1041–4.

Bryson, R. A. and T. J. Murray 1977. *Climates of hunger: mankind and the world's changing weather.* Madison, Wisc.: University of Wisconsin Press.

Bullock, T. H. 1955. Compensation for temperature in metabolism and activity of poikilotherms. *Biol Rev.* **30**, 311–42.

Bursell, E. 1964. Environmental aspects: temperature. In *The physiology of Insecta,* Vol. 1, M. Rockstein (ed.), 283–321. New York: Academic Press.

Burton, H., T. Lloyd-Evans and D. N. Weir 1970. Wryneck breeding in Scotland. *Scott. Birds* **6**, 154–6.

Butzer, K. W., G. L. Isaac, J. L. Richardson and C. Washbourn-Kamau 1972. Radiocarbon dating of East African lake levels. *Science, NY* **175**, 1069–76.

Bystrak, D. and C. S. Robbins 1977. Bird population trends detected by the North American breeding bird survey. *Pol. Ecol. Stud.* **3**, 131–143.

Cain, A. J. and J. D. Currey 1963. Area effects in *Cepaea. Phil. Trans R. Soc. B* **246**, 1–81.

Cameron, R. A. D. 1970a. The survival, weight loss and behaviour of three species of land snail in conditions of low humidity. *J. Zool. Lond.* **160**, 143–57.

Cameron, R. A. D. 1970b. The effects of temperature on the activity of three species of Helicid snail (Mollusca: Gastropoda). *J. Zool. Lond.* **160**, 303–15.

Campbell, A., B. D. Frazer, N. Gilbert, A. P. Gutierrez and M. MacKauer 1974. Temperature requirements of some aphids and their parasites. *J. Appl. Ecol.* **11**, 431–8.

Chu, Ko-chen 1963. *Development of phenology. Knowledge of it in the ancient world and laws of phenology* (in Chinese). Peking.

Clark, J. E. 1936. The history of British phenology. *Q. Jl. R. met. Soc.* **62**, 19–23.

Clark, J. E. and I. D. Margary 1930. Floral isophenes and isakairs. *Q. Jl. R. met. Soc.* **56**, 45–57.

Clark, S. C. 1969. Some effects of temperature and photoperiod on growth and floral development in three winter annuals. *New Phytol.* **68**, 1137–44.

Clench, H. K. 1966. Behavioural thermoregulation in butterflies. *Ecology* **47**, 1021–34.

Cloudsley-Thompson, J. L. 1967. *Microecology.* London: Edward Arnold.

Cloudsley-Thompson, J. L. 1971. *The temperature and water relations of reptiles.* Watford: Merrow.

Clouston, B. and K. Stansfield (eds) 1979. *After the elm . . .* London: Heinemann.

Coaker, T. H. and D. W. Wright 1963. The influence of temperature on the emergence of the cabbage root fly (*Erioischia brassicae* (Bouché)) from overwintering pupae. *Ann. Appl. Biol.* **52**, 337–43.

Coale, H. K. 1911. Enormous death rate among waterfowl near Salt Lake City, Utah, fall of 1910. *Auk* **28**, 274.

Cogley, J. G. 1979. Albedo contrast and glaciation due to continental drift. *Nature, Lond.* **279**, 712–3.

Collingwood, C. A. 1958. A survey of Irish Formicidae. *Proc. R. Ir. Acad.* **59**, 213–219.

Connell, J. H. 1961. The influence of interspecific competition and other factors on the distribution of the barnacle *Chthamalus stellatus. Ecology* **42**, 710–23.

Coope, G. R. 1970. Interpretations of Quaternary insect fossils. *A. Rev. Ent.* **15**, 97–120.

Coope, G. R. 1979. Late Cenozoic fossil Coleoptera: Evolution, Biogeography and Ecology. *A. Rev. Ecol. Syst.* **10**, 247–67.

Coope, G. R., A. Morgan and P. J. Osborne 1971. Fossil Coleoptera as indicators of climatic fluctuations during the last glaciation in Britain. *Palaeogeogr. Palaeoclim. Palaeoecol.* **10**, 87–101.

Corbet, P. S. 1957. The life-history of the emperor dragonfly *Anax imperator* Leach (Odonata: Aeshnidae). *J. Anim. Ecol.* **26**, 1–69.

Corbet, P. S., C. Longfield and N. W. Moore 1960. *Dragonflies.* London: Collins New Naturalist.

Cottam, C. 1935. The eelgrass shortage in relation to waterfowl. *Trans Am. Game Conf.* **20**, 272–9.

Cox, C. B. and P. D. Moore 1980. *Biogeography: an ecological and evolutionary approach,* 3rd edn. Oxford: Blackwell Scientific.

Craddock, J. M. 1974. Phenological indicators and past climates. *Weather* **29**, 332–43.

Cramp, S., W. R. P. Bourne and D. Saunders 1974. *The seabirds of Britain and Ireland.* London: Collins.

Crawford-Sidebotham, J. J. 1972. The influence of weather upon the activity of slugs. *Oecologia, Berl.* **9**, 141–54.

Creber, G. T. 1977. Tree rings: a natural data-storage system. *Biol Rev.* **52**, 349–83.

Creed, E. R. 1971. Melanism in the two-spot ladybird *Adalia bipunctata* in Great Britain. In *Ecological genetics and evolution,* E. R. Creed (ed.), 134–51. Oxford: Blackwell Scientific.

Crisp, D. J. 1957. Effect of low temperature on the breeding of marine animals. *Nature, Lond.* **179**, 1138–9.

Crisp, D. J. 1960. Mobility of barnacles. *Nature, Lond.* **188**, 1208–9.

Crisp, D. J. 1965. Observations on the effects of climate and weather on marine communities. In *The biological significance of climatic changes in Britain,* C. J. Johnson and L. P. Smith (eds), 63–77. London: Academic Press.

Crisp, D. J. 1964. The effect of the severe winter of 1962/63 on marine life in Britain. *J. Anim. Ecol.* **33**, 165–210.

Crisp, D. J. and A. J. Southward 1960. Recent changes in the distribution of marine organisms in north-west Europe. *Proc. 1st. int. Oceanogr. Congr. NY* **159**, 148–51.

Cushing, D. H. 1961. On the failure of the Plymouth herring fishery. *J. Mar. Biol. Assoc. UK* **41**, 799–816.

Cushing, D. H. 1976. The impact of climatic change on fish stocks in the North Atlantic. *Georgr. J.* **142**, 216–27.

Cushing, D. H. and R. R. Dickson 1976. The biological response in the sea to climatic changes. *Adv. Mar. Biol.* **14**, 1–122.

Dance, S. P. 1970. Trumpet ram's-horn snail in North Wales. *Nature Wales* **12**, 10–14.

Dansgaard, W. 1964. Stable isotopes in precipitation. *Tellus* **16**, 436–68.

Dansgaard, W., S. J. Johnsen, H. B. Clausen and C. C. Langway 1971. Climatic record revealed by the Camp Century ice core. In *The late Cenozoic glacial ages*, K. K. Turekian (ed.), 37–56. New Haven, Conn.: Yale University Press.

Davidson, J. 1944. On the relationship between temperature and the rate of development of insects at constant temperatures. *J. Anim. Ecol.* **13**, 26–38.

Davidson, J. and H. G. Andrewartha 1948a. Annual trends in a natural population of *Thrips imaginis* (Thysanoptera). *J. Anim. Ecol.* **17**, 193–9.

Davidson, J. and H. G. Andrewartha 1948b. The influence of rainfall, evaporation and atmospheric temperature on fluctuations in the size of a natural population of *Thrips imaginis* (Thysanoptera). *J. Anim. Ecol.* **17**, 200–22.

Davis, M. B. 1961. Pollen diagrams as evidence of late glacial climatic change in southern New England. *Ann. NY Acad. Sci.* **95**, 623–31.

Davis, M. B. 1963. On the theory of pollen analysis. *Am. J. Sci.* **261**, 897–912.

Delmas, R. J., J. M. Ascencio and M. Legrand 1980. Polar ice evidence that atmospheric CO_2 20 000 years BP was 50% of present. *Nature, Lond.* **284** 155–7.

Dempster, J. P. 1971. Some observations on a population of the small copper butterfly *Lycaena phlaeas* (Linnaeus) (Lep., Lycaenidae). *Entomologist's Gaz.* **22**, 199–204.

Dempster, J. P., M. L. King and K. H. Lakhani 1976. The status of the swallowtail butterfly in Britain, *Ecol. Ent.* **1**, 71–84.

Destombes, J.-P., E. R. Shephard-Thorn and J. H. Redding 1975. A buried valley system in the Strait of Dover. *Phil. Trans R. Soc. A* **279**, 243–56.

De Valera, M. 1942. A red alga new to Ireland: *Asparagopsis armata* on the west coast. *Irish Nat. J.* **8** 30–3.

De Vries, J. 1977. Histoire du climat et économie: des faits nouveaux, une interprétation différente. *Ann.: Econ. Soc., Civil.* **32**, 198–226.

Dickson, R. R. and H. H. Lamb 1972. *A review of recent hydrometeorological events in the North Atlantic Sector.* Publ. Int. Commission North-West Atlantic Fish. **8**, 35–62.

Dickson, R. R., H. H. Lamb, S.-A. Malmberg and J. M. Colebrook 1975. Climatic reversal in northern North Atlantic. *Nature, Lond.* **256**, 479–82.

Diehl, B. 1971. Productivity investigation of two types of meadows in the Vistula Valley. XII. Energy requirement in nestling and fledgling red-backed shrike (*Lanius collurio* L.) *Ekol. Polska* **19**, 235–48.

Digby, P. S. B. 1955. Factors affecting the temperature excess of insects in sunshine. *J. Exp. Biol.* **32**, 279–98.

Dixon, A. F. G. 1972. Control and significance of the seasonal development of colour forms in the sycamore aphid *Drepanosiphum platanoides* (Schr.). *J. Anim. Ecol.* **41**, 689–97.

Dixon, A. F. G. 1973. *Biology of aphids.* London: Edward Arnold.

Dixon, A. F. G. 1976. Timing of egg hatch and viability of the sycamore aphid, *Drepanosiphum platanoides* (Schr.), at bud burst of sycamore *Acer pseudoplatanus* L. *J. Anim. Ecol.* **45**, 593–603.

Dixon, P. S. 1965. Changing patterns of distribution in marine algae. In *The biological significance of climatic changes in Britain*, C. J. Johnson and L. P. Smith (eds), 109–15. London: Academic Press.

Dmi'el, R., A. Prevulotzky and A. Shkolnik 1980. Is a black coat in the desert a means of saving metabolic energy? *Nature, Lond.* **283**, 761–2.

Dobinson, H. M. and A. J. Richards 1964. The effects of the severe winter of 1962/63 on birds in Britain. *Br. Birds* **57**, 373–434.

Doornkamp, J. C. 1980. *Atlas of drought in Britain 1975–76.* London: Institute of British Geographers.

Dorsett, D. A. 1962. Preparation for flight in hawk-moths. *J. Exp. Biol.* **39**, 579–88.

Dorst, J. 1962. *The migration of birds*. London: Heinemann.

Douwes, P. 1976. Activity of *Heodes virgaureae* (Lep., Lycaenidae) in relation to air temperature, solar radiation and time of day. *Oecologia, Berl.* **22**, 287–98.

Dove, L. D. 1968. Nitrogen distribution in tomato plants during drought (*Lycopersicum esculentum* 'Manglobe'). *Phyton. Rev. Int. Bot. Exp.* **25**, 49–52.

Dunbar, M. J. 1976. Climatic change and northern development. *Arctic* **29**, 184–93.

Dunbar, M. J. and D. H. Thomson 1979. West Greenland salmon and climatic change. *Medd Grønland* **202**, 1–19.

Dunn, E. K. 1973. Changes in feeding ability of terns associated with windspeed and sea surface conditions. *Nature, Lond.* **244**, 520–1.

Dunn, J. A. and D. W. Wright 1955. Population studies of the pea aphid in East Anglia. *Bull. Ent. Res.* **46**, 369–87.

Durand, A. L. 1972. Landbirds over the North Atlantic: unpublished records 1961–65 and thoughts a decade later. *Br. Birds* **65**, 428–42.

Durango, S. 1946. Blåkråkan (*Coracias g. garrulus* L.) i Sverige. *Fågelvärld* **5**, 145–90.

Durango, S. 1950. Om klimatets inverkan på törnskatans (*Lanius collurio* L.) utbredning och levnadsmojligheter. *Fauna och Flora* **46**, 49–78.

Durango, S. 1956. Territory in the red-backed shrike *Lanius collurio. Ibis* **98**, 476–84.

Eddy, J. A. 1976. The Maunder Minimum. *Science, NY* **192**, 1189–202.

Ehrlich, P. R. and L. C. Birch 1967. The 'balance of nature' and 'population control'. *Am. Nat.* **101**, 97–107.

Ekholm, S. 1975. Fluctuations in butterfly frequency in Central Nyland. *Notulae Ent.* **55**, 65–80.

Elkins, N. 1979. Nearctic landbirds in Britain and Ireland: a meteorological analysis. *Br. Birds* **72**, 417–33.

Elson, K. G. R. 1968. Salmon disease in Scotland. *The Salmon Net* **4**, 9–17.

Elton, C. S. 1925. The dispersal of insects to Spitzbergen. *Trans Ent. Soc. Lond.* **1925** 289–299.

Elton, C. S. 1927. *Animal ecology*. London: Sidgwick & Jackson.

Emiliani, C. 1955. Pleistocene temperatures. *J. Geol.* **63**, 538–78.

Emiliani, C. 1966. Palaeotemperature analysis of Caribbean cores P 6304–8 and P 6304–9 and a generalized temperature curve for the past 415 000 years. *J. Geol.* **74**, 109–26.

Epstein, S., R. P. Sharp and A. J. Gow 1970. Antarctic ice sheet: stable isotope analysis of Byrd station cores and interhemispheric climatic implications. *Science, NY* **168**, 1570–2.

Ericson, D. B. and G. Wollin 1966. *The deep and the past*. London: Jonathan Cape.

Faegri, K. and J. Iversen 1975. *Textbook of pollen analysis*, 3rd edn. Oxford: Blackwell Scientific.

Fay, L. D. 1966. Type E botulism in Great Lakes waterbirds. *Trans 31st N. Am. Wildl. Conf.* **1966**, 139–49.

Fearn, G. M. 1973. Biological Flora of the British Isles: *Hippocrepis comosa* L. *J. Ecol.* **61**, 915–26.

Feeny, P. 1970. Seasonal changes in oak leaf tannins and nutrients as a cause of spring feeding by winter moth caterpillars. *Ecology* **51**, 565–81.

Felton, J. C. 1974. Some comments on the aculeate fauna. In *The changing flora and fauna of Britain*, D. L. Hawksworth (ed.), 399–418. London: Academic Press.

Ferguson, C. W. 1970. Dendrochronology of bristlecone pine, *Pinus aristata*. Establishment of a 7484-year chronology in the White Mountains of east-central California, USA. In *Radiocarbon variations and absolute chronology*, I.U. Olsson (ed.), 237–59. New York: Wiley.

Fischer-Piette, E. and M. Prennant 1956. Distribution des Cirrhipèdes intercotidaux d'Espagne septrentionale. *Bull. Cen. Etud. Rech. sci. Biarritz* **1**, 7–19.

Fischer-Piette, E., R. Heim and R. Lami 1932. Note préliminaire sur une maladie bactérienne des Zostères. *C. r. hedb. Séanc. Acad. Sci., Paris* **195**, 1420–2.

Flegg, J. J. M. 1972. The puffin on St Kilda 1969–1971. *Bird Study* **19**, 7–12.

Flint, R. F. and W. A. Gale 1958. Stratigraphy and radiocarbon dates at Searles Lake, California. *Am. J. Sci.* **256**, 689–714.

Ford, E. 1933. An account of the herring investigations conducted at Plymouth during the years from 1924 to 1933. *J. Mar. Biol. Assoc. UK* **19**, 305–84.

Ford, E. B. 1957. *Butterflies,* 3rd edn. London: Collins New Naturalist.

Ford, M. J. 1978a. Locomotory activity and the predation strategy of the wolf-spider *Pardosa amentata* (Clerck) (Lycosidae). *Anim. Behav.* **26**, 31–5.

Ford, M. J. 1978b. The variety of biological effects of climatic change. In *Proceedings of the Nordic symposium on climatic changes and related problems,* K. Frydendahl (ed.), 98–102. Copenhagen: Danish Meteorological Institute.

Ford, M. J. 1978c. The current colder phase could affect 'borderline' cropping. *The Grower* **89**, 746–7.

Ford, M. J. and H. H. Lamb 1976. The climate of East Anglia since historical times. In *Nature in Norfolk,* 23–8. Norwich: Jarrold for the Norfolk Naturalists' Trust.

French, R. A. and J. H. White 1960. The diamond-back moth outbreak of 1958. *Pl. Path.* **9**, 77–84.

Fritts, H. C. 1974. Relationships of ring widths in arid-site conifers to variations in monthly temperature and precipitation. *Ecol. Monogr.* **44**, 411–40.

Fritts, H. C. 1976. *Tree rings and climate.* New York: Academic Press.

Frömming, E. 1956. *Biologie der mitteleuropäischen Süsswasserschnecken.* Berlin: Duncker and Humblot.

Fry, F. E. G. 1947. *Effects of the environment on animal activity.* Publ. Ontario Fish. Res. Lab. No. 55.

Fryer, G. 1954. The trumpet ramshorn snail *Menetus (Micromenetus) dilatatus* (Gould) east of the Pennines. *Naturalist, Hull* **1954**, 86.

Gage, S. H., M. K. Mukerji and R. R. Randell 1976. A predictive model for seasonal occurrence of three grasshopper species in Saskatchewan (Orthoptera: Acrididae). *Can. Ent.* **108**, 245–53.

Gardarsson, A. and R. Moss 1970. Selection of food by Icelandic Ptarmigan in relation to its availability and nutritive value. In *Animal populations in relation to their food resources,* A. Watson (ed.), 47–69. British Ecological Society Symposium No. 10. Oxford: Blackwell Scientific.

Geiger, R. 1966. *The climate near the ground.* Cambridge, Mass.: Harvard University Press.

Gibb, J. A. 1962. L. Tinbergen's hypothesis of the role of specific search images. *Ibis* **104**, 106–11.

Gibbs, J. N. 1974. *Biology of Dutch elm disease.* Forest Record No 94. London: HMSO for Forestry Commission.

Giltner, L. T. and J. F. Couch 1930. Western duck sickness and botulism. *Science, NY* **72**, 660.

Gimingham, C. H. 1960. Biological flora of the British Isles: *Calluna vulgaris* (L.) Hull, *J. Ecol.* **48**, 455–83.

Gimingham, C. H. 1972. *Ecology of heathlands.* London: Chapman & Hall.

Glaisher, J. 1849. On the reduction of the thermometrical observations made at the apartments of the Royal Society from the years 1774–1781, and from the years 1787 to 1843. *Phil. Trans R. Soc. Lond.* **1849** 307–18.

Glaisher, J. 1850. Sequel to a paper on the reduction of the thermometrical observations made at the apartments of the Royal Society. *Phil. Trans R. Soc. Lond.* **1850**, 569–608.

Gloyne, R. W. 1972. The 'growing season' at Eskdalemuir Observatory, Dumfriesshire. *Met. Mag.* **102**, 174–8.

Godwin, H. 1975. *The history of the British Flora,* 2nd edn. Cambridge: Cambridge University Press.

Good, R. D'O. 1936. On the distribution of the lizard orchid *(Himantoglossum hircinum* Koch.). *New Phytol.* **35**, 142–70.

Goss-Custard, J. D. 1969. The winter feeding ecology of the redshank *Tringa totanus. Ibis* **111**, 338–56.

Gösswald, K. and K. Bier, 1954. Untersuchen der Casten Determination in der Gattung *Formica. Insectes Sociaux* **1**, 305–18.

Grant, K. J. 1937. An historical study of the migrations of *Celerio lineata* Fab. and *Celerio lineata livornica* Esp. (Lepidoptera). *Trans R. Ent. Soc. Lond.* **86**, 345–57.

Gray, A. J. and R. Scott 1977. Biological flora of the British Isles: *Puccinellia maritima* (Huds). *J. Ecol.* **65**, 699–716.

Gray, B. M. 1974. Early Japanese winter temperatures. *Weather* **29**, 103–7.

Gray, B. M. 1975. Japanese and European winter temperatures. *Weather* **30**, 359–68.

Greenslade, P. J. M. 1964. Pitfall trapping as a method for studying populations of Carabidae (Coleoptera.) *J. Anim. Ecol.* **33**, 301–10.

Groves, K. S., S. R. Mattingley and A. F. Tuck 1978. Increased atmospheric carbon dioxide and stratospheric ozone. *Nature, Lond.* **273** 711–5.

Haagsma, J. 1974. Etiology and epidemiology of botulism in water-fowl in the Netherlands. *Tijdschr. Diergeneesk.* **99**, 434–42.

Haagsma, J., H. J. Over, T. Smit and J. Hoekstra 1972. Botulism in waterfowl in the Netherlands in 1970. *Neth. J. Vet. Sci.* **5**, 12–23.

Hairston, N. G., F. E. Smith, and L. B. Slobodkin 1960. Community structure, population control, and competition. *Am. Nat.* **94**, 421–425.

Hardy, A. C. and P. S. Milne 1937. Insect drift over the North Sea. *Nature, Lond.* **140**, 510–1.

Hardy, A. C. and P. S. Milne 1938. Aerial drift of insects. *Nature, Lond.* **141**, 602–3.

Harper, J. L. 1967. A Darwinian approach to plant ecology. *J. Ecol.* **55**, 242–70.

Harper, J. L. 1977. *Population biology of plants.* London: Academic Press.

Harris, M. P. 1976a. The present status of the puffin in Britain and Ireland. *Br. Birds* **69**, 239–64.

Harris, M. P. 1976b. The seabirds of Shetland in 1974. *Scott. Birds* **9**, 37–68.

Harris, M. P. and S. Murray 1977. Puffins on St. Kilda. *Br. Birds* **70**, 50–65.

Harrison, C. J. O. 1961. Woodlark population and habitat. *Lond. Bird Rep.* **24**, 71–80.

Harvey, P. H. 1974. The distribution of three species of helicid snail in east Yorkshire. II. Intensive survey. *Proc. malac. Soc. Lond.* **41**, 57–64.

Haslam, S. M. 1970. The performance of *Phragmites communis* Trin. in relation to water-supply. *Ann. Bot.* **34**, 867–77.

Haslam, S. M. 1972. Biological flora of the British Isles: *Phragmites communis* Trin. *J. Ecol.* **60**, 585–610.

Hastenrath, S. 1971. On snowline depression and atmospheric circulation in the tropical Americas during the Pleistocene. *S. Afr. Geogr. J.* **53**, 53–68.

Hawke, E. L. 1953. Obituary on Major H. C. Gunton MBE, F. R. Ent. Soc. *Q. Jl. R. Met. Soc.* **62**, 19–23.

Hays, J. D., J. Imbrie, and N. J. Shackleton 1976. Variations in the Earth's orbit: pacemaker of the ice ages. *Science, NY* **194**, 1121–32.

Hearn, K. A. and M. G. Gilbert 1977. *The effects of the 1976 drought on sites of nature conservation interest in England and Wales.* Banbury: Nature Conservancy Council.

Heinrich, B. 1972. Energetics of temperature regulation and foraging in a bumblebee, *B. terricola* Kirby. *J. Comp. Physiol.* **77**, 49–64.

Heinrich, B. 1974. Thermoregulation in endothermic insects. *Science, NY* **185**, 747–56.

Hewitt, G. M. and F. M. Brown 1970. The B-chromosome system of *Myrmeleotettix maculatus*. V. A steep cline in East Anglia. *Heredity, Lond.* **25**, 363–71.

Hewitt, G. M. and B. John 1967. The B-chromosome system of *Myrmeleotettix maculatus* (Thunb.). III. The statistics. *Chromosoma* **21**, 140–62.

Hewitt, G. M. and B. John 1970. The B-chromosome system of *Myrmeleotettix maculatus* (Thunb.). IV. The dynamics. *Evolution* **24**, 169–80.

Hogg, W. H. 1965. Climatic factors and choice of site, with special reference to horticulture. In *The biological significance of climatic changes in Britain.* C. J. Johnson and L. P. Smith (eds), 141–55. London: Academic Press.

Hollstein, E. 1965. Jahrringchronologische Datierung von Eichenhölzern ohne Waldkante (Westdeutsche Eichenchronologie). *Bónn. Jb.* **165**, 1–27.

Hopkins, J. W. 1968. Protein content of Western Canadian hard red spring wheat in relation to some environmental factors. *Agric. Met.* **5**, 411–431.

Howarth, T. G. 1973. *South's British Butterflies.* London: Warne.

Howe, R. W. 1967. Temperature effects on embryonic development in insects. *A. Rev. Ent.* **12**, 15–42.

Huber, B. and V. Giertz-Siebenlist 1969. Unsere tausendjährige Eichen-Jahrringchronologie durchschnittlich 57 (10–150) – fach belegt. *Sber. Akad. Wiss. Wien I: Biol., Mineral., Erdkunde verwand. Wiss.* **178**, 37–42.

Hughes, I. G. 1975. Changing altitude and habitat preferences of two species of wood-ant (*Formica rufa* and *F. lugubris*) in North Wales and Salop. *Trans R. Ent. Soc. Lond.* **127**, 227–39.

Hugh-Jones, M. E. and P. B. Wright 1970. Studies on the 1967–8 foot-and-mouth disease epidemic: the relation of weather to the spread of the disease. *J. Hyg., Camb.* **68**, 253–71.

Hunt, O. D. 1965. Status and conservation of the large blue butterfly, *Maculinea arion* L. In *The conservation of invertebrates*, E. Duffey and M. G. Morris (eds), 35–44. Monks Wood Experimental Station Symposium No. 1.

Hurst, G. W. 1963. Small mottled willow moth in southern England, 1962. *Met. Mag.* **92**, 308–12.

Hurst, G. W. 1965. *Laphygma exigua* immigrations into the British Isles, 1947–1963. *Int. J. Biomet.* **9**, 21–8.

Hurst, G. W. 1968. Foot and mouth disease, the possibility of continental sources of the virus in England in epidemics of October 1967 and several other years. *Vet. Rec.* **81**, 610–7.

Hurst, G. W. 1969. Meteorological aspects of insect migrations. *Endeavour* **28**, 77–81.

Hurst, G. W. 1970. Can the Colorado beetle fly from France to England? *Entomologist's Mon. Mag.* **105**, 269–72.

Hutchins, L. W. 1947. The bases for temperature zonation in geographical distribution. *Ecol Monogr.* **11**, 325–35.

Hutchinson, G. E. 1957. Concluding remarks. *Cold Spring Harbour Symp. Quant. Biol.* **22**, 415–27.

Imbrie, J. and J. Z. Imbrie 1980. Modelling the climatic response to orbital variations. *Science, NY* **207**, 943–53.

Ingram, M. J., D. J. Underhill and T. M. L. Wigley 1978. Historical climatology. *Nature, Lond.* **276**, 329–34.

International Bird Census Committee 1969. Recommendations for an international standard for a mapping method in bird census work. *Bird Study* **16**, 249–55.

Isaksen, I. S. A., E. Hesstvedt and F. Stordal 1980. Influence of stratospheric cooling from CO_2 on the ozone layer. *Nature, Lond.* **283**, 189–91.

Iversen, J. 1944. *Viscum, Hedera* and *Ilex* as climate indicators. *Geol. Fören. Stockholm Förhandl.* **66**, 463–483.

Ives, W. G. H. 1974. *Information report NOR-X-118.* Northern Forest Research Centre: Edmonton, Alberta.

Jackes, Å. D. and A. Watson 1975. Winter whitening of Scottish mountain hares (*Lepus timidus scoticus*) in relation to daylength, temperature and snow-lie. *J. Zool. Lond.* **176**, 403–9.

Jackson, H. C. 1978. Low May sunshine as a possible factor in the decline of the sand lizard (*Lacerta agilis*) in North-West England. *Biol Conserv.* **13**, 1–12.

Järvinen, O. and R. A. Väisänen, 1979. Climatic changes, habitat changes, and competition: dynamics of geographical overlap in two pairs of congeneric bird species in Finland. *Oikos* **33**, 261–71.

Jeffree, E. P. 1960. Some long-term means from 'The Phenological Reports' (1891–1948) of the Royal Meteorological Society. *Q. Jl. R. Met. Soc.* **86**, 95–103.

Johnson, C. G. 1969. *Migration and dispersal of insects by flight.* London: Methuen.

Jones, P. D. (ed.) 1976. *Climate Monitor* **5** (1). Norwich: Climatic Research Unit, University of East Anglia.

Kalela, O. 1949. Changes in geographic ranges in the avifauna of northern and central Europe in relation to recent changes in climate. *Bird-Banding* **20**, 77–103.

Kalmbach, E. R. 1930. Western duck sickness produced experimentally. *Science, NY* **72**, 658–60.

Kemble, A. R. and H. T. MacPherson 1954. Liberation of amino acids in perennial ryegrass during wilting. *Biochem. J.* **58**, 46–50.

Kemp, S. 1938. Oceanography and the fluctuation in the abundance of marine animals. *Rep. Br. Assoc.* **1938**, 85–101.

Kennett, J. P. and R. C. Thunell 1975. Global increase in Quaternary explosive volcanism. *Science, NY* **187**, 497–503.

Kerney, M. P., E. H. Brown and T. J. Chandler, 1964. The late-glacial and post-glacial history of the chalk escarpment near Brook, Kent. *Phil. Trans R. Soc. B* **248**, 135–204.

Kettlewell, H. B. D. 1958. A survey of the frequencies of *Biston betularia* (L.) (Lep.) and its melanic forms in Great Britain. *Heredity, Lond.* **12**, 51–72.

Kettlewell, H. B. D. 1961. The phenomenon of industrial melanism in Lepidoptera. *A. Rev. Ent.* **6**, 245–62.

Kettlewell, H. B. D. 1973. *The evolution of melanism.* Oxford: Clarendon Press.

Keymer, I. F., G. R. Smith, T. A. Roberts, S. I. Heaney and D. J. Hibberd 1972. Botulism as a factor in waterfowl mortality at St James' Park, London. *Vet. Rec.* **90**, 111–4.

King, J. W. 1973. Solar radiation changes and the weather. *Nature, Lond.* **245**, 443–6.

Kington, J. A. 1974. An application of phenological data to historical climatology. *Weather* **29**, 320–8.

Koch, L. 1945. The east Greenland ice. *Medd Grønland* **130**, (3).

Köppen, W. 1923. *Die Klimate der Erde.* Berlin and Leipzig: Bomträger.

Köppen, W. 1931. *Grundriss der Klimakunde.* Berlin and Leipzig: Bomträger.

Kristjansson, L. 1969. The ice drifts back to Iceland. *New Scient.* **41**, 508–9.

Krogh, A. 1914. The quantitative relation between temperature and standard metabolism in animals. *Int. Z. Phys.-chem. Biol.* **1** 491–508.

Krogh, A. and E. Zeuthen, 1941. The mechanism of flight preparation in some insects. *J. Exp. Biol.* **18**, 1–10.

Kukla, G. J. and H. J. Kukla, 1974. Increased surface albedo in the northern hemisphere. *Science, NY* **183**, 709–14.

Lack, D. 1954a. *The natural regulation of animal numbers.* Oxford: Clarendon Press.

Lack, D. 1954b. The stability of the heron population. *Br. Birds* **47**, 111–21.

Lack, D. 1966. *Population studies of birds.* Oxford: Clarendon Press.

Lack, D. and E. Lack, 1951. The breeding biology of the swift *Apus apus. Ibis* **93**, 501–46.

La Marche, V. C. Jr 1974. Palaeoclimatic inferences from long tree-ring records. *Science, NY* **183**, 1043–8.

La Marche, V. C. Jr and T. P. Harlan 1973. Accuracy of tree-ring dating of bristlecone pine for calibration of the radiocarbon time scale. *J. Geophys. Res.* **78**, 8849–58.

Lamb, H. H. 1963. What can we find out about the trend of our climate? *Weather* **18**, 194–216.

Lamb, H. H. 1964a. Trees and climatic history in Scotland. *Q. Jl. R. Met. Soc.* **90**, 382–94.

Lamb, H. H. 1964b. Climatic changes and variations in the atmospheric and oceanic circulations. *Geol. Rundschau* **54**, 486–504.

Lamb, H. H. 1965. The early mediaeval warm epoch and its sequel. *Palaeogeogr., Palaeoclim., Palaeoecol.* **1**, 13–37.

Lamb, H. H. 1966. Climate in the 1960s: changes in the world's wind circulation reflected in prevailing temperatures, rainfall patterns and the levels of the African lakes. *Geogr. J.* **132**, 183–212.

Lamb, H. H. 1967. Britain's changing climate. *Geogr. J.* **133**, 445–68.

Lamb, H. H. 1970. Volcanic dust in the atmosphere. *Phil. Trans R. Soc. A* **266**, 425–533.

Lamb, H. H. 1971. Volcanic activity and climate. *Palaeogeogr. Palaeoclim. Palaeoecol.* **10**, 203–30.

Lamb, H. H. 1972a. British Isles weather types and a register of the daily sequence of circulation patterns 1861–1971. *Geophys. Mem.* **116**.

Lamb, H. H. 1972b. *Climate present, past and future.* Vol. 1: *Fundamentals and climate now.* London: Methuen.

Lamb, H. H. 1974. Climate, vegetation and forest limits in early civilized times. *Phil. Trans R. Soc. Lond. A* **276**, 195–230.

Lamb, H. H. 1977a. *Climate present, past and future.* Vol. 2: *Climatic history and the future.* London: Methuen.

Lamb, H. H. 1977b. *Understanding climatic change and its relevance to the world food problem.* CRU Research Publication No 5. Norwich: Climatic Research Unit, University of East Anglia.

Lamb, H. H. 1978. The variability of climate: observation and understanding. In *Proceedings of the Nordic symposium on climatic change and related problems,* K. Frydendahl (ed.), 116–44. Copenhagen: Danish Meteorological Institute.

Lamb, H. H. (in press). *Climate, history and the modern world.* London: Methuen.

Lamb, H. H. and A. I. Johnson 1966. Secular variations of the atmospheric circulation since 1750. *Geophys. Mem.* **110**.

Landsberg, H. E., C. S. Yu and L. Huang 1968. *Preliminary reconstruction of a long time series of climatic data for the eastern United States.* Tech. Note BN-571. Baltimore: University of Maryland (Institute of Fluid Dynamics).

Landsberg, J. J. 1979. From bud to bursting blossom: weather and the apple crop. *Weather* **34**, 394–407.

Langford, T. E. 1972. A comparative assessment of thermal effects in some British and North American rivers. In *River ecology and man,* R. Oglesby, C. Carlson and J. McCann (eds), 319–51. New York: Academic Press.

Lantz, L. A. 1927. Quelques observations nouvelles sur l'herpetologie des Pyrenées centrale. *Rev. Hist. nat. appliq.* **8**, 54–61.

Leith, H. (ed.) 1974. *Phenology and seasonality modelling.* Ecological Studies No. 8. Berlin: Springer-Verlag.

Leivestad, H. and I. P. Muniz 1976. Fish kills at low pH in a Norwegian river. *Nature, Lond.* **259**, 391–2.

Le Roy Ladurie, E. 1972. *Times of feast, times of famine: a history of climate since the year 1000.* London: George Allen & Unwin.

Likens, G. E., R. F. Wright, J. N. Galloway and T. J. Butler 1979. Acid rain. *Scient. Am.* **241**, (4), 39–47.

Lloyd, C. S., G. J. Thomas, J. W. MacDonald, E. D. Borland, K. Standing and J. L. Smart 1976. Wild bird mortality caused by botulism in Britain, 1975. *Biol Conserv.* **10**, 119–29.

Locket, G. H. and A. F. Millidge 1951. *British spiders* Vol. 1. London: Ray Society.

Locket, G. H., A. F. Millidge and P. Merrett 1974. *British spiders,* Vol 3. London: Ray Society.

Lockley, R. M. 1953. *Puffins.* London: Dent.

Longfield, C. 1948. A vast immigration of dragonflies into the south coast of Co. Cork. *Ir. Nat. J.* **9**, 133–41.

Lucas, J. A. W. 1950. The algae transported on drifting objects and washed ashore on the Netherlands coast. *Blumea* **6**, 527–43.

Lumby, J. R. and G. T. Atkinson 1929. On the unusual mortality amongst fish during March and April 1929 in the North Sea. *J. Cons. Int. Explor. Mer* **4**, 309–32.

Lusis, J. J. 1961. On the biological meaning of colour polymorphism of ladybeetle *Adalia bipunctata* L. *Latvijas Ent.* **4**, 3–29.

Luther, H. H. 1951. Verbreitung und Ökologie der höheren Wasserpflanzen im Brackwasser der Ekenäs-Gegend in Südfinnland. II. Spezieller Teil. *Acta bot. fenn.* **50**, 1–370.

Lyall, I. T. 1970. Recent trends in spring weather. *Weather* **26**, 163–5.

Macan, T. T. 1974. Freshwater invertebrates. In *The changing flora and fauna of Britain,* D. L. Hawksworth (ed.), 143–55. London: Academic Press.

Machta, L. 1972. Mauna Loa and global trends in air quality. *Bull. Am. Met. Soc.* **53**, 402–20.

Manabe, S. and R. Wetherald, 1975. The effects of doubling the CO_2 concentration on the climate of a general circulation model. *J. Atmos. Sci.* **32**, 3–15.

Manley, G. 1953. The mean temperature of central England, 1698–1952. *Q. Jl. R. Met. Soc.* **79**, 242–61.

Manley, G. 1974. Central England temperatures: monthly means 1659 to 1973. *Q. Jl. R. Met. Soc.* **100**, 389–405.

Manley, G. 1975. 1684: the coldest winter in the English instrumental record. *Weather* **30**, 382–8.

Mann, K. H. 1958. Occurrence of an exotic oligochaete *Branchiura sowerbyi* Beddard, 1892, in the River Thames. *Nature, Lond.* **182**, 732.

Margary, I. D. 1926. The Marsham phenological record in Norfolk, 1736–1925, and some others. *Q. Jl. R. Met. Soc.* **52**, 27–54.

Marriner, T. F. 1927. Observations on the life history of Subcoccinella 24-punctata. Entomologist's Mon. Mag. **63**, 118–23.

Marshall, J. A. 1974. The British Orthoptera since 1800. In *The changing flora and fauna of Britain,* D. L. Hawksworth (ed.), 307–22. London: Academic Press.

Marsham, R. 1789. Indications of spring. *Phil. Trans R. Soc.* **79**, 154.

Martin, M. H. 1968. Conditions affecting the distribution of *Mercurialis perennis* L. in certain Cambridgeshire woodlands. *J. Ecol.* **56**, 777–93.

Martin, W. R. and A. C. Kohler 1965. *Variations in recruitment of cod* (Gadus morhua L.) *in southern ICNAF waters as related to environmental changes.* Sp. Publ. Int. Commission North-west Atlantic Fish. **6**, 833–46.

Mathias, J. H. 1971. *The comparative ecologies of two species of amphibia* (Bufo bufo and Bufo calamita) *on the Ainsdale sand dunes national nature reserve.* Ph.D. Thesis, University of Manchester.

Mattson, W. J. and N. D. Addy 1975. Phytophagous insects as regulators of forest primary production. *Science, NY,* **190,** 515–22.

McIntyre, A. 1967. Coccoliths as palaeoclimatic indicators of Pleistocene glaciation. *Science, NY,* **158,** 1314.

McLean, I., N. Carter and A. Watt 1977. Pests out of control? *New Scient.* **76,** 74–5.

McLean, R. C. 1938. Carriers of foot and mouth disease. *Nature, Lond.* **141,** 828.

McNalty, A. 1943. Indigenous malaria in England. *Nature, Lond.* **151,** 440–2.

Mead, C. J. 1973. Movements of British raptors. *Bird Study* **20,** 259–86.

Mech, L. D. 1966. The wolves of Isle Royale. *Fauna Nat. Pks U.S.* **7,** 1–210.

Mellanby. K. 1939. Low temperature and insect activity. *Proc. R. Soc.* **127,** 473–87.

Merikallio, E. 1951. Der Einfluss der letzten Wärmeperiode (1930–49) auf die Vogelfauna Nordfinnlands. *Proc. Xth int. Ornithol. Congr., Uppsala* **1951** 484–93.

Milankovitch, M. 1930. Mathematische Klimalehre und astronomische Theorie der Klimaschwankungen. In *Handbuch der Klimatologie* I, Teil A, W. Köppen and R. Geiger (eds). Berlin: Bornträger.

Miller, R. S. 1967. Pattern and process in competition. *Adv. Ecol. Res.* **4,** 1–74.

Mitchell, J. M. 1961. Recent secular changes of global temperature. *Ann. NY Acad. Sci.* **95,** 235–50.

Moore, H. B. 1936. The biology of *Balanus balanoides.* V. Distribution in the Plymouth area. *J. Mar. Biol. Assoc. UK* **20,** 701–16.

Moore, N. W. 1953. Population density in adult dragonflies (Odonata-Anisoptera). *J. Anim. Ecol.* **22,** 344–59.

Moore, N. W. 1962. The heaths of Dorset and their conservation. *J. Ecol.* **50,** 369–91.

Moore, N. W. 1975. Butterfly transects in a linear habitat 1964–73. *Entomologist's Gaz.* **26,** 71–8.

Morris, R. F. 1963. *The dynamics of epidemic spruce budworm populations.* Mem. Ent. Soc. Can. No. 31.

Mörzer Bruijns, M. D. 1955. The brent goose (*Branta bernicla*) on Terschelling (Netherlands). *Ardea* **43,** 261–71.

Mounce, I. and W. W. Diehl, 1934. A new *Ophiobolus* on eelgrass. *Can. J. Res.* **11,** 242–6.

Muggleton, J. 1973. Some aspects of the history and ecology of blue butterflies in the Cotswolds. *Proc. Br. Ent. Nat. Hist. Soc.* **6,** 77–84.

Muggleton, J. 1974. Dates of appearance of *Maculinea arion* (Linnaeus) (Lep. Lycaenidae) adults in Gloucestershire 1858–1960. *Entomologist's Gaz.* **25,** 239–44.

Muggleton, J., D. Lonsdale and B. R. Benham 1975. Melanism in *Adalia bipunctata* L. (Col. Coccinellidae) and its relationship to atmospheric pollution. *J. Appl. Ecol.* **12,** 451–63.

Müller, K. 1953. *Geschichte des Badischen Weinbaus.* Lahr in Baden: von Moritz Schauenburg.

Münchberg, P. 1931. Zur Biologie der Odonatagenera *Brachytron* Evans und *Aeschna* Fbr. *Z. Morph. Ökol. Tiere* **20,** 172–232.

Murdoch, W. W. 1966. 'Community structure, population control and competition' – a critique. *Am. Nat.* **100,** 219–26.

Nicholas, F. J. and J. Glasspoole 1931. General monthly rainfall over England and Wales 1727–1931. *Br. Rainfall* **1931,** 299–306.

Nicholson, A. J. 1933. The balance of animal populations. *J. Anim. Ecol.* **2,** 132–78.

Nicholson, A. J. 1954. An outline of the dynamics of animal populations. *Aust. J. Zool.* **2,** 9–65.

Norgaard, E. 1951. On the ecology of two Lycosid spiders *Pirata piraticus* and *Lycosa pullata* from a Danish sphagnum bog. *Oikos* 3, 1–21.

Odum, H. T. 1967. Biological circuits and the marine systems of Texas. In *Pollution and marine ecology*, T. A. Olson and F. J. Burgess (eds), 99–157. New York: Wiley.

Ogilvie, M. A. and A. K. M. St Joseph 1976. Dark-bellied brent geese in Britain and Europe 1955–76. *Br. Birds* 69, 422–39.

Owen, D. F. 1976. Ladybird, ladybird fly away home. *New Scient.* 71, 686–7.

Panigel, M. 1956. Contribution à l'étude de l'ovoviparité chez les reptiles: gestation et parturition chez le lézard vivipare *Zootoca vivipara*. *Annls. Sci. Nat. (Zool.)* 18, 569–668.

Park, T. 1954. Experimental studies of interspecific competition. II. Temperature, humidity and competition in two species of *Tribolium*. *Physiol. Zoöl.* 27, 177–238.

Park, T. 1962. Beetles, competition and populations. *Science, NY* 138, 1369–75.

Patel, B. and D. J. Crisp, 1960. The influence of temperature on the breeding and moulting activities of some warm water species of operculate barnacles. *J. Mar. Biol. Assoc. UK* 39, 667–80.

Peakall, D. B. 1962. The past and present status of the red-backed shrike in Great Britain. *Bird Study* 9, 198–216.

Perring, F. H. 1965. The advance and retreat of the British flora. In *The biological significance of climatic changes in Britain*, C. J. Johnson and L. P. Smith (eds), 51–9. London: Academic Press.

Perring, F. H. 1974. Changes in our native vascular plant flora. *The changing flora and fauna of Britain*, D. L. Hawksworth (ed), 7–25. London: Academic Press.

Perring, F. H. and S. M. Walters 1962. *Atlas of the British Flora*. London: Nelson.

Perry, A. H. 1971. Changes in position and intensity of major Northern Hemisphere 'centres of action'. *Weather* 26, 268–70.

Peterken, G. F. and P. S. Lloyd 1967. Biological flora of the British Isles: *Ilex Aquifolium* L. *J. Ecol.* 55, 841–58.

Phillipson, J. 1962. Respirometry and the study of energy turnover in natural systems with particular reference to harvestspiders (Phalangida). *Oikos* 13, 311–22.

Pigott, C. D. 1968. Biological flora of the British Isles: *Cirsium acaulon* (L.) Scop. *J. Ecol.* 56, 597–612.

Pigott, C. D. 1970. The response of plants to climate and climatic change. In *The flora of a changing Britain*, F. H. Perring (ed.), 32–44. Botanical Society of the British Isles Reports No. 11. Faringdon, Berks.: Classey.

Pigott, C. D. 1975. Experimental studies on the influence of climate on the geographical distribution of plants. *Weather* 30, 82–90.

Pivarova, Z. I. 1968. The long-term variation of the intensity of solar radiation according to the observations of actinometric stations (in Russian). *Glav. Geofiz. Obs., Trudy* 233, 17–37.

Pollard, E. 1975. Aspects of the ecology of *Helix pomatia* L. *J. Anim. Ecol.* 44, 305–29.

Pollard, E. 1977. A method for assessing changes in the abundance of butterflies. *Biol Conserv.* 12, 115–34.

Pollard, E. 1979a. A national scheme for monitoring the abundance of butterflies: the first three years. *Proc. Br. Ent. Nat. Hist. Soc.* 12, 77–90.

Pollard, E. 1979b. Population ecology and change in range of the white admiral butterfly *Ladoga camilla* L. in England. *Ecol. Ent.* 4, 61–74.

Pollard, E., D. O. Elias, M. J. Skelton and J. A. Thomas 1975. A method of assessing the abundance of butterflies in Monks Wood National Nature Reserve in 1973. *Entomologist's Gaz.* 26, 79–88.

Potts, G. R. 1969. Partridge survival project. *Rep. Game Res. Assoc.* 8, 14–17.

Powell, H. T. 1957. Studies in the genus *Fucus* L. III. Distribution and ecology of forms of *Fucus distichus* L. emend. Powell in Britain and Ireland. *J. Mar. Biol. Assoc. UK* **36**, 663–93.

Prestt, I., A. S. Cooke and K. F. Corbett 1974. British amphibians and reptiles. In *The changing flora and fauna of Britain,* D. L. Hawksworth, (ed.) 229–54. London: Academic Press.

Rackham, O. 1975. *Hayley Wood: its history and ecology.* Cambridge: Cambridge and Isle of Ely Naturalists' Trust.

Rainey, R. C. 1951. Weather and movements of locust swarms: a new hypothesis. *Nature, Lond.* **168**, 1057–60.

Rainey, R. C. 1973. Airborne pests and the atmospheric environment. *Weather* **28**, 224–39.

Ranwell, D. S. and B. M. Downing 1959. Brent goose (*Branta bernicla* (L.)) winter feeding pattern and *Zostera* resources at Scott Head Island, Norfolk. *Anim. Behav.* **7**, 42–56.

Rao, K. P. and T. H. Bullock 1954. Q_{12} as a function of size and habitat temperature in poikilotherms. *Am. Nat.* **88**, 33–44.

Rasmussen, E. 1973. Systematics and ecology of the Isefjord marine fauna (Denmark). *Ophelia* **11**, 1–507.

Ratcliffe, D. A. 1968. An ecological account of Atlantic bryophytes in the British Isles. *New Phytol.* **67**, 365–439.

Ratcliffe, R. A. S., J. Weller and P. Collison 1978. Variability in the frequency of unusual weather over approximately the last century. *Q. Jl. R. met. Soc.* **104**, 243–56.

Renn, C. E. 1934. Wasting disease of *Zostera* in American waters. *Nature, Lond.* **134**, 416.

Renn, C. E. 1936. The wasting disease of *Zostera marina. Biol. Bull. Mar. Biol. Lab. Woods Hole* **70**, 148–58.

Richards, A. M., R. D. Pope and V. F. Eastop 1976. Observations on the biology of *Subcoccinella 24-punctata* (L.) in southern England. *Ecol. Ent.* **1**, 201–207.

Richards, O. W. and N. Waloff 1954. *Studies on the biology and population dynamics of British grasshoppers.* Anti-locust Bulletin No. 17.

Richardson, A. M. M. 1974. Differential climatic selection in natural population of land snail *Cepaea nemoralis. Nature, Lond.* **247**, 572–3.

Riehl, H. and J. Meitin 1980. Discharge of the River Nile: a barometer of short-period climatic variation. *Science, NY* **206**, 1178–9.

Robbins, C. S. and W. T. Van Velzen, 1974. Progress report on the North American Breeding Bird Survey. *Acta ornithol.* **14**, 170–91.

Robert, A. 1958. *Les Libellules (Odonates).* Neuchâtel: Delachaux et Niestlé.

Roberts, W. O. and H. Lansford 1979. *The climate mandate.* San Francisco: W. H. Freeman.

Robock, A. 1979. The 'Little Ice Age': Northern hemisphere average observations and model calculations. *Science, NY* **206**, 1402–4.

Rossby, C.-G. 1939. Relation between variations in the intensity of the zonal circulation of the atmosphere and the displacements of the semi-permanent centres of action. *J. Mar. Res.* **2**, 38–55.

Rossby, C.-G. 1941. The scientific basis of modern meteorology. *US Ybk. Agric.* **1941**, 599–655.

Royama, T. 1970. Factors governing the hunting behaviour and selection of food by the great tit (*Parus major* L.). *J. Anim. Ecol.* **39**, 619–68.

Ruddiman, W. F. and L. K. Glover 1974. Counterclockwise circulation in the North Atlantic subpolar gyre during the Quaternary. In *Proceedings of the international conference on mapping the atmospheric and other climatic parameters at the time of the last glacial maximum about 17000 years ago,* Climatic Research Publication No. 2 49–51. Norwich: Climatic Research Unit, University of East Anglia.

Rudy, H. 1925. *Die Wanderheuschrecke* (Locusta migratoria *L.,* Phasa migratoria *L. und* Phasa danica *L.*). *Beiträge zu einer Monographie.* Frieberg: Sonderbeilage zur Badischen Blätter fur Schädlings-bekämpfung.

Russell, F. S. 1935. On the value of certain planktonic animals as indicators of water movements in the English Channel and the North Sea. *J. Mar. Biol. Assoc. UK* **20**, 309–32.

Russell, F. S. 1973. A summary of the observations of the occurrence of planktonic stages of fish, Plymouth 1924–72. *J. Mar. Biol. Assoc. UK* **53**, 347–55.

Russell, F. S., A. J. Southward, G. T. Boalch and E. I. Butler 1971. Changes in the biological conditions in the English Channel off Plymouth during the last half-century. *Nature, Lond.* **234**, 468–70.

Salomonsen, F. 1948. The distribution of birds and the recent climatic change in the North Atlantic area. *Dansk Orn. Foren. Tidsskr.* **42**, 85–89.

Salomonsen, F. 1951. The immigration and breeding of the fieldfare in Greenland. *Proc. Xth Int. Orn. Congr.,* Uppsala 1950, pp. 241–4.

Salomonsen, F. 1958. The present status of the brent goose in western Europe. *Vidensk. Meddr Dansk Naturh. Foren.* **120**, 43–80.

Sargent, F. and S. W. Tromp 1964. *A survey of human biometeorology.* WMO Technical Note 65. Geneva: World Meteorological Organization.

Satchell, J. E. 1965. Extinctions and invasions – some case histories and conclusions. In *The conservation of invertebrates,* E. Duffey and M. G. Morris (eds), 19–28. Monks Wood Experimental Station Symposium No. 1.

Satchell, J. E. and C. A. Collingwood 1955. The wood ants of the English Lake District. *North Western Nat.* **3**, 23.

Saville, A. 1963. A decline in the yield of Scottish estuarine spring spawning herring fisheries. *Rapp. Cons. Explor. Mer* **154**, 215–9.

Schenk, J. 1924. Der Zug der Waldschnepfe in Europe. *Aquila* **30**, 26–120.

Schoener, T. W. 1971. Theory of feeding strategies. *A. Rev. Ecol. Syst.* **2**, 369–404.

Segner, W. P., C. F. Schmidt and J. K. Boltz 1971. Minimal growth temperature, sodium chloride tolerance, pH sensitivity and toxin production of marine and terrestrial strains of *Clostridium botulinum* type C. *Appl. Microbiol.* **22**, 1025–9.

Sellers, A. and A. J. Meadows, 1975. Long term variations in the albedo and surface temperature of the Earth. *Nature, Lond.* **254**, 44.

Setchell, W. A. 1929. Morphological and phenological notes on *Zostera marina* L. *Univ. Calif. Pubs Bot.* **114**, 289–452.

Seward, D. 1979. *Monks and wine.* London: Mitchell Beazley.

Shackleton, N. J. and N. D. Opdyke 1976. Oxygen isotope and palaeomagnetic stratigraphy of Pacific core V28–239, Late Pliocene to Latest Pleistocene. *Geol Soc. Am. Mem.* **145**, 449–64.

Sharrock, J. T. R. 1974. The changing status of breeding birds in Britain and Ireland. In *The changing flora and fauna of Britain,* D. L. Hawksworth (ed.), 203–20. London: Academic Press.

Shaw, M. W. 1962. The diamond-back moth migration of 1958. *Weather* **17**, 221.

Simpson, A. C. 1953. Some observations on the mortality of fish and the distribution of plankton in the southern North Sea during the cold winter of 1946–47. *J. Cons. perm. int. Explor. Mer* **19**, 150–77.

Singh, T. N., L. G. Paleg and D. Aspinall 1973. Stress metabolism. I. Nitrogen metabolism and growth in the barley plant during water stress. *Aust. J. Biol. Sci.* **26**, 45–56.

Smith, C. V. 1964. Some evidence of the windborne spread of fowl pest. *Met. Mag.* **93**, 257–262.

Smith, F. R. 1972. Rarity Records Committee Report on rare birds in Great Britain in 1971. *Br. Birds* **65**, 322–54.

Smith, G. R. 1976. Botulism in waterfowl. *Wildfowl* **27**, 129–38.

Smith, L. P. 1956. Potato blight forecasting by 90% humidity criteria. *Pl. Path.* **5**, 83–7.

Smith, L. P. and M. E. Hugh-Jones 1969. The effects of wind and rain on the spread of foot-and-mouth disease. *Nature, Lond.* **223**, 712–5.

Smith, L. P. and J. Walker 1966. Simplified weather criteria for potato blight infection periods. *Pl. Path.* **15**, 113–6.

Solomon, M. E. 1949. The natural control of animal populations. *J. Anim. Ecol.* **18**, 1–35.

Somero, G. N. 1969. Enzymic mechanisms of temperature compensation. *Am. Nat.* **103**, 517–30.

Sørensen, T. 1953. A revision of the Greenland species of *Puccinellia*. *Meddr Grønland* **136**, 1–179.

Southern H. N. 1970. The natural control of a population of tawny owls (*Strix aluco*). *J. Zool. Lond.* **162**, 197–285.

Southward, A. J. 1955. On the behaviour of barnacles. I. The relation of cirral and other activities to temperature. *J. Mar. Biol. Assoc. UK* **34**, 403–22.

Southward, A. J. 1963. The distribution of some plankton animals in the English Channel and approaches. III. Theories about long-term biological changes, including fish. *J. Mar. Biol. Assoc. UK* **43**, 1–29.

Southward, A. J. 1967. Recent changes in abundance of intertidal barnacles: a possible effect of climatic deterioration. *J. Mar. Biol. Assoc. UK* **47**, 81–95.

Southward, A. J. 1974. Changes in the plankton community of the Western English Channel. *Nature, Lond.* **249**, 180–1.

Southward, A. J. 1980. The Western English Channel – an inconstant ecosystem? *Nature, Lond.* **285**, 361–6.

Southward, A. J. and D. J. Crisp 1954. Recent changes in the distribution of the intertidal barnacles *Chthamalus stellatus* (Poli) and *Balanus balanoides* (L.) in the British Isles. *J. Anim. Ecol.* **23**, 163–77.

Southward, A. J. and D. J. Crisp 1956. Fluctuations in the distribution and abundance of intertidal barnacles. *J. Mar. Biol. Assoc. UK* **35**, 211–29.

Southward, A. J. and A. D. Mattacola 1980. Occurrence of Norway pout *Trisopterus esmarki* (Nilsson) and blue whiting *Micromesistius poutassou* (Risso) in the western English Channel off Plymouth. *J. Mar. Biol. Assoc. UK.* **60**, 39–44.

Southward, A. J., E. I. Butler and L. Pennycuick, 1975. Recent cyclic changes in climate and in abundance of marine life. *Nature, Lond.* **253**, 714–7.

Southwell, T. 1875. On Mr Marsham's 'Indications of Spring' *Trans Norfolk Norwich Nat. Soc.* **2**, 31–45.

Southwell, T. 1901. On Mr Marsham's 'Indications of Spring' *Trans Norfolk Norwich Nat. Soc.* **7**, 246–54.

Southwood, T. R. E. 1960. The flight activity of Heteroptera. *Trans R. Ent. Soc. Lond.* **112**, 173–200.

Southwood, T. R. E. and D. J. Cross 1969. The ecology of the partridge. III. Breeding success and the abundance of insects in natural habitats. *J. Anim. Ecol.* **38**, 497–509.

Sparks, B. W. and R. G. West 1972. *The Ice Age in Britain.* London: Methuen.

Speerschneider, C. I. H. 1915. *Om Isforholdene i danske Farvande i aeldre og nyere Tid: aarene 690–1860.* Meddelelser Nr 2. Copenhagen: Danish Meteorological Institute.

Spellerberg, I. F. 1973. Critical minimum temperatures of reptiles. In *Effects of temperature on ectothermic organisms,* W. Wieser (ed.), 239–47. Berlin: Springer-Verlag.

Spellerberg, I. F. 1977. Adaptations of reptiles to cold. In *Morphology and biology of reptiles,* A. d'A. Bellairs and C. B. Cox (eds), 261–85. London: Linnean Society.

Spence, D. H. N. 1964. The macrophytic vegetation of freshwater lochs, swamps and associated fens. In *The vegetation of Scotland,* J. H. Burnett (ed.), 306–425. Edinburgh: Oliver & Boyd.

Spooner, G. M. 1963. On causes of the decline of *Maculinea arion* L. (Lep., Lycaenidae) in Britain. *Entomologist* **96**, 199–210.

Stafford, J. 1971. Heron populations of England and Wales 1928–70. *Bird Study* **18**, 218–21.

Steuerwald, B. A. and D. L. Clark 1972. *Globigerina pachyderma* in Pleistocene and Recent Arctic Ocean sediment. *J. Palaeontol.* **46**, 573–80.

Stuart, M. R. and H. T. Fuller 1968. Mycological aspects of diseased Atlantic salmon. *Nature, Lond.* **217**, 90–2.

Svardson, G. and S. Durango 1950. Spring weather and population fluctuations. *Proc. int. Orn. Congr.* **10**, 636–44.

Tanasijevic, No. 1958. Zur Morphologie und Biologie des Luzernemarienkafers *Subcoccinella vigintiquatuorpunctata* L. (Coleoptera: Coccinellidae). *Beitr. Ent.* **8**, 23–78.

Taylor, L. R. 1963. Analysis of the effect of temperature on insects in flight. *J. Anim. Ecol.* **32**, 99–117.

Taylor, L. R., R. A. French and E. D. M. Macaulay 1973. Low-altitude migration and diurnal flight periodicity; the importance of *Plusia gamma* L. (Lepidoptera: Plusiidae). *J. Anim. Ecol.* **42**, 751–60.

Taylor, R. H. R. 1948. The distribution of reptiles and Amphibia in the British Isles, with notes on species recently introduced. *Br. J. Herpetol.* **1**, 1–38.

Templeman, W. 1965. *Mass mortalities of marine fishes in the Newfoundland area presumably due to low temperature.* Sp. Publ. Int. Commission North-West Atlantic Fish. **6**, 523–33.

Thomas, J. A. 1976. *The black hairstreak. Conservation report.* Cambridge: Institute of Terrestrial Ecology.

Thornthwaite, C. W. 1933. The climates of the Earth. *Geogr. Rev.* **23**, 433–40.

Thornthwaite, C. W. 1948. An approach to a rational classification of climate. *Geogr. Rev.* **38**, 55–94.

Thoroddsen, Th. 1916. *Árferði á Íslandi í pusund ár* (Climate of Iceland in 1000 years). Copenhagen: Hið. islenska fraeðafélag.

Tiensuu, L. 1934. The Odonata of Sortavala parish. *Ann. Soc. Zool.-bot. Vanamo, Helsingfors* **14**, 75–114.

Tromp, S. W. (ed.) 1963. *Medical biometeorology.* Amsterdam: Elsevier.

Tubbs, C. R. 1967. Numbers of Dartford warblers in England during 1962–66. *Br. Birds* **60**, 87–9.

Tubbs, C. R. 1977. Wildfowl and waders in Langstone Harbour. *Br. Birds* **70**, 177–99.

Turrill, W. B. 1937. The black knapweed and its use in phenology. *Q. Jl. R. Met. Soc.* **63**, 79–81.

Tutin, T. G. 1938. The autecology of *Zostera marina* in relation to its wasting disease. *New Phytol.* **37**, 50–71.

United Nations 1977. *Desertification: an overview.* Conference on Desertification. Document A/CONF.74/1/Rev.1. Nairobi: United Nations.

Urey, H. C. 1947. The thermodynamic properties of isotopic substances. *J. Chem. Soc.* **152**, 190–219.

Vaadia, Y., F. L. Raney, and R. M. Hagen 1961. Plant water deficits and physiological processes. *A. Rev. Pl. Physiol.* **12**, 265–92.

Vaino, I. 1932. Zur Verbreitung und Biologie der Kreutzotter, *Vipera berus,* in Finnland. *Annls Soc. Zool. Bot. Fenn.* **12**, 1–19.

Van Zinderen Bakker, E. M. 1969. Intimations on Quaternary palaeoecology of Africa. *Acta bot. Neerl.* **18**, 230–9.

Varley, G. C. 1947. The natural control of population balance in the knapweed gall-fly (*Urophora jaceana*). *J. Anim. Ecol.* **16**, 139–87.

Varley, G. C. and G. R. Gradwell 1958. Balance in insect populations. *Proc. 10th Int. Congr. Ent, Montreal* **1956**, (2), 619–24.

Varley, G. C. and G. R. Gradwell 1963. The interpretation of change and stability in insect populations. *Proc. R. Ent. Soc. Lond. C* **27**, 52–7.

Varley, G. C. and G. R. Gradwell 1968. Population models for the winter moth. In *Insect abundance,* T. R. E. Southwood (ed.), 132–42. Oxford: Blackwell Scientific.

Vernekar, A. D. 1968. *Research on theory of climate. Vol. 2: Long-period global variations of incoming radiation.* Hartford, Conn.: Travelers Research Center.

Volsøe, H. 1944. Structure and seasonal variation of the male reproductive organs of *Vipera berus* (L.). *Spolia zool. Mus. haun.* **5**, 1–157.

Walès-Smith, B. G. 1971. Monthly and annual totals of rainfall representative of Kew, Surrey from 1697–1970. *Met. Mag.* **100**, 345–62.

Walker, D. and R. G. West (eds) 1970. *Studies in the vegetational history of the British Isles. Essays in honour of Harry Godwin.* Cambridge: Cambridge University Press.

Wallén, C. C. 1953. The variability of summer temperature in Sweden and its connection with changes in the general circulation. *Tellus* **5**, 157–78.

Watson, A. 1963. The effect of climate on the colour changes of mountain hares (*Lepus timidus scoticus*) in Scotland. *Proc. Zool. Soc. Lond.* **141**, 823–35.

Watt, W. B. 1968. Adaptive significance of pigment polymorphisms in *Colias* butterflies. I. Variation of melanin pigment in relation to thermoregulation. *Evolution* **22**, 437–58.

Webb, N. R. and L. E. Haskins 1980. An ecological survey of heathlands in the Poole Basin, Dorset, England in 1978. *Biol Conserv.* **17**, 281–96.

Webb, T. and R. A. Bryson 1972. Late- and Postglacial climatic change in the northern midwest USA: quantitative estimates derived from fossil pollen spectra by multivariate statistical analysis. *Quaternary Res.* **2**, 70–115.

Webley, D. 1964. Slug activity in relation to weather. *Ann. Appl. Biol.* **53**, 407–14.

Wegener, A. 1912. *Die Entstehung der Kontinente und Ozeane.* Vieweg: Braunschweig.

Wells, G. P. 1943. The water relations of snails and slugs. III Factors determining activity in *Helix pomatia* L. *J. Exp. Biol.* **20**, 79–87.

Wells, T. C. E. and D. M. Barling 1971. Biological Flora of the British Isles: *Pulsatilla vulgaris* Mill. (*Anemone pulsatilla* L.). *J. Ecol.* **59**, 275–92.

Wendland, W. M. and R. A. Bryson 1970. Atmospheric dustiness, man and climatic change. *Biol Conserv.* **2**, 125–8.

West, R. G. 1968. *Pleistocene geology and biology.* London: Longman.

White, M. G. 1954. The house longhorn beetle *Hylotrupes bajulus* L. (Col. Cerambycidae) in Great Britain. *Forestry* **27**, 31–40.

White, T. C. R. 1969. An index to measure weather-induced stress of trees associated with outbreaks of psyllids in Australia. *Ecology* **50**, 905–9.

White, T. C. R. 1974. A hypothesis to explain outbreaks of looper caterpillars, with special reference to populations of *Selidosema suavis* in a plantation of *Pinus radiata* in New Zealand. *Oecologia, Berl.* **16**, 279–301.

White, T. C. R. 1976. Weather, food and plagues of locusts. *Oecologia, Berl.* **22**, 119–34.

Whittow, G. C. (ed.) 1970. *Comparative physiology of thermoregulation,* Vol. 1: *Invertebrates and non-mammalian vertebrates.* New York: Academic Press.

Whittow, G. C. (ed.) 1971. *Comparative physiology of thermoregulation,* Vol. 2: *Mammals.* New York: Academic Press.

Wieser, W. (ed.) 1973. *Effects of temperature on ectothermic organisms.* Berlin: Springer-Verlag.

Wigley, T. M. L., P. D. Jones and P. M. Kelly 1980. Scenario for a warm, high-CO_2 world. *Nature, Lond.* **283**, 17–21.

Williams, C. B. 1930. *The migration of butterflies.* Edinburgh: Oliver & Boyd.

Williams, C. B. 1940. An analysis of four years captures of insects in a light trap. Part II. The effect of weather conditions on insect activity; and the estimation and forecasting of changes in the insect population. *Trans R. Ent. Soc. Lond.* **90**, 228–306.

Williams, C. B. 1958. *Insect migration.* London: Collins New Naturalist.

Williams, C. B. 1961. Studies in the effect of weather conditions on the activity and abundance of insect populations. *Phil. Trans R. Soc. B* **244**, 331–78.

Williamson, K. 1952. Migrational drift in Britain in autumn 1951. *Scott. Nat.* **64**, 2–18.

Williamson, K. 1958. Bergmann's rule and obligatory overseas migration. *Br. Birds* **51**, 209–32.

Williamson, K. 1969. Weather systems and bird movements. *Q. Jl. R. Met. Soc.* **95**, 414–23.

Williamson, K. 1970. *The Common Bird Census as a device for monitoring population levels.* Bull. Ecol. Res. Comm. No. 9. Lund, Sweden.

Williamson, K. 1974. New bird species admitted to the British and Irish lists since 1800. In *The changing flora and fauna of Britain,* D. L. Hawksworth (ed.), 221–7. London: Academic Press.

Williamson, K. 1976. Recent climatic influences on the status and distribution of some British birds. *Weather* **31**, 362–84.

Williamson, K. and R. C. Holmes 1964. Methods and preliminary results of the Common Bird Census 1962–63. *Bird Study* **11**, 240–56.

Winstanley, D. 1973a. Recent rainfall trends in Africa, the Middle East and India. *Nature, Lond.* **243**, 464–5.

Winstanley, D. 1973b. Rainfall patterns and general atmospheric circulation. *Nature, Lond.* **245**, 190–4.

Winstanley, D., R. Spencer and K. Williamson 1974. Where have all the whitethroats gone? *Bird Study* **21**, 1–14.

Witherby, H. F. 1928. A trans-Atlantic passage of lapwings. *Br. Birds* **22**, 6–13.

Wohlschlag, D. E. 1960. Metabolism of an antarctic fish and the phenomenon of cold adaptation. *Ecology* **41**, 287–92.

World Meteorological Association, 1976. *An evaluation of climate and water resources for development of agriculture in the Sudano-Sahelian zone of West Africa.* Special Environmental Report No. 9. Geneva: WMO.

Woodhead, P. M. J. and A. D. Woodhead, 1959. The effects of low temperatures on the physiology and distribution of the cod, *Gadus morhua* L., in the Barents Sea. *Proc. Zool. Soc. Lond.* **133**, 181–99.

Woodward, F. I. 1975. The climatic control of the altitudinal distribution of *Sedum rosea* (L.) Scop. and *S. telephium* L. II. The analysis of plant growth in controlled environments. *New Phytol.* **74**, 335–48.

Woodward, F. I. and C. D. Pigott 1975. The climatic control of the altitudinal distribution of *Sedum rosea* (L.) Scop. and *S. telephium* L. I. Field observations. *New Phytol.* **74**, 323–34.

Wright, A. E. 1942. *Plebejus argus* Linnaeus race *masseyi* (Tutt) in North Lancashire and South Westmorland. *Entomologist* **75**, 7–13.

Wynne-Edwards, V. C. 1962. *Animal dispersion in relation to social behaviour.* Edinburgh: Oliver & Boyd.

Yoshino, M. M. 1974. Agricultural climatology in Japan. In *Agricultural meteorology of Japan,* Y. Mihara (ed.), 11–40. Tokyo: University of Tokyo Press.

Young, E. L. 1938. *Labrinthula* on Pacific coast eel-grass. *Can. J. Res.* **16**, 115–7.

Ziegler, P. 1969. *The Black Death.* London: Collins.

Species index

Page numbers printed in italics refer to the location of text figures.

General index

Page numbers printed in italics refer to the location of text figures. Text section numbers are printed in bold type.